The Foundations of Immunology and their Pertinence to Medicine

by Peter Bretscher

 FriesenPress

Suite 300 - 990 Fort St
Victoria, BC, Canada, V8V 3K2
www.friesenpress.com

Copyright © 2016 by Peter Bretscher
First Edition — 2016

ISBN
978-1-4602-9655-4 (Hardcover)
978-1-4602-9656-1 (Paperback)
978-1-4602-9657-8 (eBook)

1. Medical, Immunology

Distributed to the trade by The Ingram Book Company

Do not let yourself be tainted with barren scepticism.

Louis Pasteur

Table of Contents

Preface

I have written this book with the hope of influencing immunologists and of providing non-specialists, medical students and clinicians, interested in immunology and its relationship to medicine, with an insight into the contemporary scene.

I begin with the question of whether there is a highly developed scientific method resulting in certain knowledge, or whether science is just the sophisticated expression of everyday reasoning and exploring? It might seem surprising to raise this as the first question in the book. However, I believe its consideration will provide both immunologists and the interested non-immunologist with confidence that they can understand the ensuing discussion of how the immune system operates and of its relationship to medicine. Moreover, a consideration of this question has both inspired the science I do and provided me with the motivation to write this book.

Science has led to knowledge of which we can be surprisingly confident. For example, there are prescriptive manuals on how to calculate the strength and stability of different kinds of bridge. These manuals are the result of past scientific endeavors. However, they reflect technical information rather than current scientific enquiry. When science is so successful that it can be reliably used for practical purposes, we recognize that it has given birth to technology. Large manuals are written. When consulting such a manual, few question the basis of the knowledge on which the strength and stability of different types of bridge are calculated. Such manuals are usually only accessible to specialists.

Current bridge manuals are based on Newtonian physics and so, from a modern perspective, are strictly speaking invalid. No modern physicist thinks about the physical world primarily in terms of Newtonian concepts. This example illustrates the difference between practical utility and live scientific enquiry, where in principle nothing should be taken for granted, and the struggle to have valid foundations is paramount.

The history of science teaches us that the framework in which our scientific knowledge is cast needs to be radically changed under certain, usually paradoxical, circumstances. This is most apparent when scientific theories have evolved to be successful in their accounting of a wide body of natural phenomena, but are also found to be strikingly deficient in some other respect; for example, being incompatible with some well established observations that are expected to be explicable within the theory's mandate. Alternatively, the theory may be unsatisfactory by being in conflict with some principle appealing to investigators. The history of physics of the last century is replete with such scenarios. However, anyone familiar with the history of other scientific disciplines, immunology included, can see the same kind of pattern. The persistent reader will recognize such scenarios in the pages that follow.

A moral can be drawn from these considerations. No scientific method can be employed to mechanically crank out certain knowledge. If there is no such method, how is science different from everyday reasoning and exploration? I argue that the most significant science is usually done within the context of a contemporary culture. We have learnt from the giants of science and of philosophy just how subtle and significant reasoning and the exploration of ideas can be. I argue that such an appreciation is an essential component of scientific culture, with the result that subtle arguments are seriously considered and paid attention to by the scientific community. This culture also reflects the virtue of making and testing predictions of different frameworks to assess their utility in understanding nature. To relive the past experiences of science, by becoming familiar with its history, is empowering. It is appropriate to refer to such collective empowerment as the culture supporting scientific endeavors.

Different scientific styles lead to different types of contribution. Some contributions, made within the context of a generally accepted framework, explore the framework's utility to account for the breadth of known and of new observations. To follow in detail the arguments of such contributions often requires considerable training in the discipline, and extensive knowledge of the field. The extreme form of such contributions is like employing a prescriptive manual for building bridges. Such considerations are certainly not readily accessible to the non-specialist, and can even be daunting to the specialist with a different area of expertise.

At the other end of the spectrum are contributions that question the nature of the currently accepted solutions to basic issues of the field, and that propose novel solutions. These high-risk potential contributions typically either fall flat with time or can result in a dramatic reorientation of research. This has happened in the last fifty years of immunology a few times, as I shall relate.

It is perhaps not surprising that basic issues are rather few in number, because they are foundational. The proposed solutions to these issues provide the axioms or framework on which explanations are developed. Indeed, foundational studies require the ability of the investigator to identify what issues are truly basic and whose solutions are consequently seminal.

A consideration of foundational ideas does not require a detailed knowledge of all the observations that the framework can explain, nor of the technical developments on which such explanations are based. Foundational ideas can usually be simply expressed. Moreover, an appreciation of just one clear and simple paradox can show a framework to be unsatisfactory, requiring a re-examination of its foundations. Thus, the considerations involved in addressing foundational ideas are open to understanding by a non-specialist with only moderate knowledge.

In the recent past, by which I mean the period of two hundred to one hundred years ago, western science was much more foundational than today's because it was less developed and so more accessible to the non-specialist. We are in the era cursed with overspecialization. Charles Darwin's foundational *Origin of Species*, published in 1859, was accessible to the broadly educated individual. This was a time when gifted individuals could attain a valid grasp of diverse aspects of the culture in which they lived.

This book describes how foundational concepts of immunology have evolved over the last two and a half centuries. It traces how, in the late 1700s, Jenner established a safe means of vaccinating against small pox, and how this success in public health led in the 1800s to the establishment of two new sciences, immunology and the study of infectious diseases. These advances led to a characterization of immune responses in the late 1800s and early 1900s. The Clonal Selection Theory, developed in the 1950s, was formulated to explain how some of these major characteristics could be realized at a mechanistic level. It is no accident that this formulation occurred at roughly the same time that molecular biology was becoming established. This new science had a decisive impact.

The Clonal Selection Theory inspired an explosion of studies in the field for the next three decades. Later investigations tended to examine the minutiae of the system and led immunologists to view the immune system as extremely complex. This assessment was fostered by an exponential increase in the number of immunologists and their publications.

This book is an attempt to transcend contemporary overspecialization by returning to an examination of foundational concepts. I anticipate that this feature will make the book accessible to interested clinicians and non-specialists. Most today consider the immune system to be highly complex. I believe that a focus on foundational concepts reveals the immune system to be sophisticated. By this, I mean that its limited complexity can be understood to serve physiological needs, and so the need for this complexity becomes understandable; these features are then seen as sophistication on the part of Mother Nature and of evolution. I shall deal with some controversial foundational concepts, and will sometimes argue for ones that have not found favor or have been abandoned by the immunological community at large.

I wish to make a final comment on style. I have striven to achieve accessibility, brevity, and clarity. The clarity I seek here is at the conceptual level. I do not, for the most part, attempt to justify concepts in terms of extensive observations that support them, or by comprehensively delineating observations that conflict with opposing concepts. I employ observations more to illustrate concepts than attempt their rigorous justification. This might be regarded as constituting more of a poetic than of a scientific approach. Circumstances allow me to be comfortable with writing a book in this style. I have recently justified most of the concepts elaborated upon here on a more documented and observational basis in my more technical book, *Rediscovering the Immune System as an Integrated Organ*, which is more than twice the length of the present volume. The few references here are chosen for their conceptual contributions to the field. I refer herein to my more technical book simply as *Rediscovering*.

I have particularly enjoyed writing this book. It is a pleasure to be free to focus on what I perceive are the basic concepts, to strive for clarity and brevity, and to try to provide an account that is accessible to the general reader. Indeed, on completing the writing of *Rediscovering*, I was surprised by the spontaneous wish and release of energy to write another immunological book, whose aim is to appeal to the reader's intuition and analytical ability, unburdened by the need for extensive observational justification.

It is my pleasure to express my gratitude to Zoltan Nagy, William Albritton and Henry Tabel for a careful reading of the manuscript and providing me with their helpful comments, and to Juliane Deubner for help with the Figures. It is also my pleasure to express my gratitude to Calliopi Havele, my wife, without whose support this book would not have seen the light of day.

A note on technical terms

My teaching experience has led me to try to develop ways of minimizing the barriers that technical terms present in achieving a valid understanding. This book has a glossary at the end, but it does not provide definitions of terms used. Instead, it notes the page(s) in which the term is defined and its use shown in context. The term, when first used and its meaning explained, is ***bolded and italicized***. I believe it is critical to use terms as precisely as possible to foster clear discussion and understanding. However, precision is only possible in a context, and so definitions can become somewhat elaborate, circular, and so not absolute. I hope this way of defining terms in context will facilitate the reader's understanding.

Chapter 1

How the immune system was discovered and its attributes recognized

Circumstances leading to the exploitation of immunity and to the science of immunology

Voltaire, the French philosopher and writer, wrote in the middle of the 1700s of his travels to China and the Middle East. Smallpox was, at this time, a prevalent and deadly disease in Europe. It is estimated that sixty of every hundred Englishmen were infected, of which twenty died and twenty bore the everlasting scars that could follow infection. Voltaire reported on practices in China and the Middle East whereby protection could be achieved against smallpox. Material was harvested from the crusts of the pocks of an infected individual, and this material was administered by a prescribed method to a healthy individual, with the aim of protecting them from subsequent disease. This process led to the death of one in a hundred of these individuals deliberately exposed in this manner[1].

Milkmaids, according to folk knowledge in England at this time, were protected against smallpox through exposure to cows suffering from cowpox, a disease similar to the human variety. In the late 1700s, Edward Jenner started systematically vaccinating against smallpox by exposing individuals to material obtained from cowpox lesions. He scarified the skin before exposing the individual to this material. This process resulted in reliable protection against smallpox, with no untoward consequences.

This landmark of western medicine was followed about fifty years later by the studies of the German, Robert Koch, and the Frenchman, Louis Pasteur. Their work over a number of years provided a context for understanding Jenner's success. They demonstrated that some diseases are caused by infectious agents. This demonstration required Koch and Pasteur to isolate the causative pathogen, to find conditions under which they could grow the pathogen in the laboratory, and then to show that infection of an animal with the organism resulted in the symptoms characteristic of the disease.

One animal model used by Pasteur was chicken cholera. This bacterial disease results in various symptoms, including violent diarrhea. Pasteur had found conditions under which he could grow the responsible bacteria. On returning from a summer vacation, he found a plate on which the bacteria had grown and that he had accidentally left on his laboratory bench. Pasteur discovered that these bacteria, when injected into chickens, would no longer cause disease. The bacteria had lost their *virulence*. However, infection with these bacteria could protect against a subsequent challenge of virulent bacteria. The bacteria had, over the summer holidays, become ***attenuated***. Following this chance finding, Pasteur developed different means of culturing pathogens under conditions less than optimal for their growth, resulting in their attenuation. An attenuated pathogen could be used to protect against the corresponding infectious disease. Pasteur called this process ***vaccination*** in honor of Jenner's discovery with cowpox virus, also called vaccinia virus. The name vaccinia is derived from vacca, the Latin for cow.

Two further findings made in the late 1800s were seminal and can be seen as the culmination of the discoveries that led to the establishment of two new and related sciences, immunology and the study of infectious diseases. First, Roux and Yersin reported in 1888 that the bacteria-free supernatants of cultures of diphtheria bacilli could, when administered to an animal, cause the symptoms of the disease. This finding led to the recognition that some bacterial pathogens cause disease by their production and secretion of *toxins.* In time, it was found that toxins could be chemically or physically treated in such a way that they were no longer toxic. Such modified non-toxic molecules are called *toxoids.* They can be used to vaccinate individuals against a normally pathogenic challenge of the toxin. The treatment thus "attenuates" the toxin.

The second seminal finding arose from an attempt to characterize the nature of the protection provided by vaccination. Von Behring and Kitasato attempted to determine if there are protective molecules in the blood of immune animals. They drew blood from an animal immune to tetanus toxin, allowed it to clot, and harvested the honey-colored, cell-free serum. They reported in 1890 that a naïve animal, given serum derived from an immune animal, is resistant to a normally lethal challenge of tetanus toxin, see Figure 1.

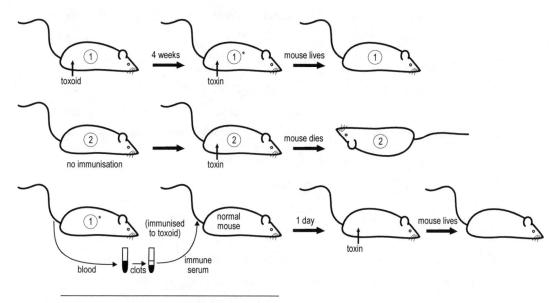

Figure 1. Passive transfer of humoral immunity

This demonstration led to the conclusion that components in the humor, the non-cellular component of blood, were responsible for protection. These protective molecules were called **antibodies**, and their presence is referred to as **humoral immunity**. Substances recognized by antibodies are called **antigens**.

These investigations led to an explosion of important discoveries. Immunology as a science was born.

The immune system's four attributes

Innate and immune defense mechanisms

All forms of life, from single cells to multi-cellular organisms, have mechanisms of defense to protect them from the environment and foreign invaders. Immune systems are a unique form of defense only found in vertebrates. We refer to the evolutionary older mechanisms of defense as **innate defense**. Vertebrates have both innate and immune defense mechanisms.

Four attributes of immune systems distinguish them from the systems of innate defense. I first outline the nature of these four attributes because understanding this is essential to appreciate the unique properties of immune systems. This recognition will provide a context for analyzing how these features are realized. The consequent understanding will provide a conceptual foundation to devise strategies for the prevention and treatment of clinical conditions in the five areas of medicine related to the immune system.

The immune system's adaptability

Innate mechanisms of defense are present either at the time of an insult or are mobilized within minutes. These mechanisms are said to be **constitutive**, that is, part of the constitution of the organism as always present. The expression of the defense mechanisms of the immune system is, in contrast, regulated. The immune system is said to be *adaptable*. This adaptability is evident in three distinct forms of regulation of the expression of immune effector mechanisms.

Firstly, it takes days if not weeks to develop effective immunity against a foreign invader such as a pathogen. Secondly, the Greek Thucydides, the first recognized western historian, noted in about 500 BCE a different form of immunological memory. He recorded that individuals who had survived a previous epidemic were resistant to becoming ill during a similar epidemic, often occurring decades later. He reported that those who had previously survived the plague could tend the sick with impunity. This observation reflects a prevalent finding. The immune systems of animals and people usually respond much more rapidly and with greater intensity on a second than a first infection by the same pathogen. It is said that the *secondary immune response* generated upon a *secondary infection* is greater than the *primary immune response* generated upon a *primary infection*. This form of adaptability is different from that associated with a state of immunity. It reflects an enhanced ability to generate a potent state of immunity. The secondary immune response is more rapid and larger than the primary response. This memory response is the basis of most vaccination, as we shall shortly see.

A third form of adaption is different from the previous two. The first two reflect positive memory in that they reflect a state of immunity that takes time to develop, or an enhanced ability to generate such immunity. It turns out that previous exposure to an antigen can reduce the immune response to a subsequent exposure to this same antigen. We shall later consider the physiological significance of these adaptive modes that negatively regulate, or ablate, immune responses.

The immune system's specificity and universality

It was shown in the late 1800s that antibodies are fairly specific. When honey colored and transparent, cell-free serum is obtained from an animal immune to tetanus toxin and is added to a transparent solution of this toxin, the mixture goes cloudy. This precipitation is due to interactions between the antibody and the toxin molecules. This phenomenon is referred to as the ***precipitin reaction***. When anti-tetanus toxin antibody is added to diphtheria toxin, no precipitate forms. Thus, the interaction of antibody with antigen expresses some specificity.

Landsteiner, in the early decades of the 1900s, found a means to raise antibodies to small, organic molecules whose structures had been elucidated in the late 1800s. A molecule able to induce the formation of antibody is said to be ***immunogenic***, capable of generating an immune response. Landsteiner showed that when he chemically coupled a small molecule, the size of a benzene ring or two, to an immunogenic protein and immunized rabbits with this conjugate, he could raise antibodies able to bind this small molecule. The small molecule is called a ***hapten*** and the large immunogenic molecule, to which the hapten is coupled to raise anti-hapten antibody, is usually referred to as a ***carrier***. Thus, anti-hapten antibody can be raised by immunizing with a ***hapten-carrier conjugate***.

Landsteiner also examined whether these anti-hapten antibodies could bind to other small molecules whose structure was closely related to the hapten. He found that most often the antibodies were less able to bind the related molecule. Landsteiner's observations led to the recognition of the highly specific nature of antibody/antigen interactions. An everyday analogy is employed to appreciate this ***specificity***. It is likened to the fact that keys usually only open their corresponding locks.

A second aspect of the antibody response became apparent from Landsteiner's work. Antibody could be raised to virtually any small foreign molecule that organic chemists devised. This remarkable ability of the immune system is referred to as ***universality***. It appeared that the immune system was able to respond to the unexpected and unanticipated. An everyday analogy illustrates how remarkable this attribute is. Consider a key shop that has keys to all locks now existing and to all possible locks, some of which have not yet been invented. We will later see how universality is achieved.

The exquisite specificity of antibodies and the universality of the immune system must mean that the immune system has the capacity to produce an incredible number of distinct antibody molecules. It is estimated that humans and vertebrates can each produce many billions!

Self-nonself discrimination

All defense systems have an ability to preferentially attack foreign invaders rather than self cells or molecules. This attribute is referred to as self-nonself discrimination. The mechanisms by which innate defense and immune systems achieve self-nonself discrimination are radically different.

Self-nonself discrimination by innate defense mechanisms

It is evident that the ability to distinguish *self* from *nonself* has to rely on properties that distinguish self from nonself cells and molecules.

There are receptors in non-vertebrates, as well as in vertebrates, called *pattern recognition receptors (PRR)*, which can bind to *pathogen-associated molecular patterns (PAMPs)*, present on infectious pathogens, and often also on their benign counterparts, but not present on self cells or molecules. Consider an insect as an example of a non-vertebrate with only innate mechanisms of defense. The different PAMPs it encounters can be regarded as flags characteristic of different classes of invaders, such as viruses, bacteria, or protozoa. The interaction of the insect's PRR with a PAMP initiates an attack by the insect upon the PAMP-bearing invader.

This kind of self-nonself discrimination is realized because the PRRs bind to structures present on foreign but not on self cells or molecules. The PRR are nearly always proteins. The particular proteins an organism produces are determined by its genes. Genes thus define the PRRs, and the PRRs recognize nonself but not self, so the ability to discriminate nonself from self is *germline encoded*, in other words, specified by the organism's genes.

Self-nonself discriminatiom by the immune system

Consider how self-nonself discrimination might be achieved by the immune system. An attribute of the immune system is its universality—the ability to respond to virtually all foreign molecules. This means that an individual grafted with a kidney from his non-identical brother can reject it, as his brother has different genes from his. In the luck of the draw that occurs at conception, the grafted individual could have inherited the genes that make his brother's kidney foreign. In this case, these antigens would have been self. This example illustrates that an individual has the ability to respond to something that could have been self. We can therefore conclude that we have the intrinsic ability to respond to self-antigens and so, to achieve self-nonself discrimination, this intrinsic ability must be ablated or somehow held in check. This tolerance towards self-antigens is not germ-line encoded, but the result of an **adaptive process**. The nature of the process of self-nonself discrimination by the immune system is thus radically different from the mechanism of innate defense.

There must be properties distinguishing self from foreign antigens if the immune system is to reliably respond to these antigens differently. Burnet and Fenner suggested in the late 1940s that the property that distinguishes self from foreign antigens was their first appearance early in development.[2] Support for this idea came from the studies of Hasek and Medawar and their colleagues in the early 1950s. It was found that the deliberate exposure of developing or new-born animals to a foreign antigen could ablate the ability of the animal as an immunologically mature individual from making an immune response to the antigen. It thus appeared that early exposure to a foreign antigen in the life history of an animal could result in the animal's immune system regarding this foreign antigen as a self-antigen. This idea is referred to as the **Historical Postulate**.[2] This acquired unresponsiveness to an antigen is a form of negative adaption on the part of the immune system, as the ability to respond to self-antigens is ablated.

Immune class regulation

The recognition that there are distinct classes of immunity grew out of Koch's attempts to develop a treatment for tuberculosis in the late 1800s.

He had isolated from tissues of tuberculosis patients the causative pathogen *Mycobacterium tuberculosis,* and found conditions under which he could grow these bacteria in the lab. Koch tried to stimulate the patients' immune system by injecting them with a protein preparation obtained from *M tuberculosis* called **purified protein derivative (PPD).** Such treatment led to inconsistent results; in some cases it appeared to lead to an improvement in the patient's symptoms and in others to a rapid deterioration of the patient's condition and sometimes even to death. Koch had to abandon this experimental treatment.

However, he made a seminal discovery during these studies. He observed a particular kind of inflammation on injecting his patients with purified protein derivative. He saw a swelling at the site of injection that first became apparent at twelve hours post-injection and peaked between twenty-four and forty-eight hours. This inflammatory response is referred to as *delayed type hypersensitivity (DTH)*, distinguishing it from acute inflammation. *Acute inflammation* is observed within minutes following the injection of antigen into the skin of an individual allergic to the antigen, as we shall shortly see. This latter inflammation is referred to as reflecting *immediate hypersensitivity*.

Studies in animals demonstrated that DTH is a highly specific reaction, similar to the specificity of the antibody/antigen interaction. Studies also showed that, in contrast to humoral immunity against a toxin, a state of DTH could not be conferred upon a naïve animal by giving it serum from an immune animal. It was surmised that cells are needed to mediate the DTH reaction. It was not possible to test this idea until the 1940s when inbred strains of various experimental animals, such as mice and guinea pigs, were developed. Mice and guinea pigs of the same sex and same strain are genetically identical, and are said to be *syngeneic*, meaning having the same genes. In this case, cells can be transferred between them without the recipient making an immune response against the donated cells. With the existence of inbred strains, it was possible to show that a naïve animal would express DTH if given cells from a syngeneic animal that had been immunized to express DTH. Thus, DTH is mediated by cells and is referred to as a form of **cell-mediated immunity**.

It might be thought that the immune system would direct all the weapons in its arsenal against a foreign invader. Diverse observations in both animals and people show this not to be the case. The immune system has a decision-making mechanism that determines whether a cell-mediated or humoral response is generated. We refer collectively to all the processes that determine the class of immunity induced as **immune class regulation**. Understanding the basis of such regulation is central to devising strategies of immunological intervention in many areas of medicine related to the immune system.

The significance of immune class regulation: recruiting different mechanisms of innate defense

We have noted that immune systems first appeared in vertebrates. The animals from which they arose had innate mechanisms of defense. The immune system exploits these evolutionarily older mechanisms. I outline four such mechanisms as a prelude to illustrating how the immune system commandeers them.

Single cell amoebae have the ability to ingest some other single cell organisms and then destroy them. They use the resulting small molecules to build their own larger molecules. This process is appropriately likened to eating and is called **phagocytosis** from the Greek *phago*, meaning I eat. All multicellular organisms have specialized cells with similar functions, and these defensive cells are called **phagocytes**.

A second form of innate defense is mediated by a group of molecules called **complement**. Complement has diverse functions, one of which is to insert a donut-shaped assembly of molecules into the membranes of invading cells, including those of bacteria. Large molecules can pass through the doughnuts' holes, leading to the bacterium's death. Figure 2 shows an electron microscopic photograph of a bacterial membrane with holes made by complement.

A third mechanism of innate defense is responsible for the inflammation that occurs when we get a substantial scratch on our skin. The redness around the scratch forms within one minute and the scratch will, if substantial, swell along its length within several minutes. Nerve cells close to the skin's surface, which respond to pressure, release a substance when the scratch is made. This substance activates mast cells, resident close by, to release several substances, one of which is called histamine. Histamine in turn causes nearby blood vessels to dilate and become leaky, leading to inflammation along the length of the scratch. This inflammation results in fluid containing protective molecules and cells, including phagocytes, to accumulate at the site of injury and deal with any invaders. This process is called *acute inflammation*, to distinguish it from delayed-type hypersensitivity, which is a different form of inflammation.

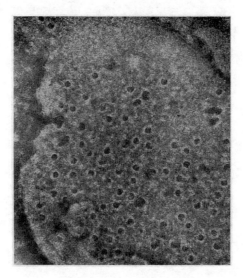

Figure 2. Holes in a bacterial membrane produced by complement

Lastly, cells infected by viruses trigger the production of *interferon*. This is secreted by the infected cells and induces neighboring cells to produce anti-viral proteins that interfere with viral replication, thereby protecting them.

The structure of antibodies

A knowledge of the structure of antibodies is essential to understand how they commandeer innate mechanisms of defense to attack an invader, an understanding that is central to explaining the full significance of immune class regulation.

Antibodies are proteins, and all proteins are made up from polypeptide chains. The prototypical antibody unit consists of two identical light and two identical heavy chains, and thus has two identical halves, as shown Figure 3. The antibody chains themselves are made up of distinct *domains*, represented in the figure as contiguous white and black regions. Light chains have two domains, while some heavy chains have four and others have five. The first domain of the light and the first of the heavy chain are called the variable domains, as there is a lot of variability between the corresponding domains of different antibody chains. These two variable domains are close together in space and interact to form the binding site of the antibody that interacts with antigen. Thus, the prototypical antibody unit shown in Figure 3 can bind two antigen molecules, meaning it is divalent.

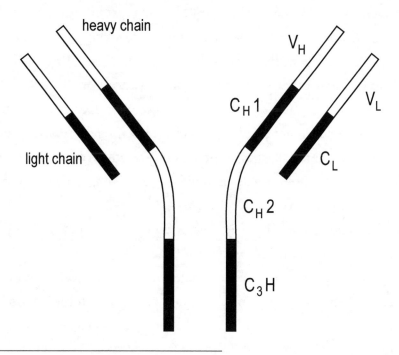

Figure 3. The prototypical unit of an antibody molecule

There are seven types of heavy chain in people, as defined by their different constant domains, and seven classes/subclasses of antibody. The nature of the constant domains of an antibody molecule's heavy chain defines the class or subclass of antibody to which the molecule belongs. The classes are IgM, IgA, IgE, and IgG, the latter containing the IgG_1, IgG_2, IgG_3, and IgG_4 subclasses.

Molecules belonging to the IgM class consist of five prototypical units, and so have a valency of 10 (2x5). The IgE and IgG molecules all contain just one unit, and so are divalent. IgA molecules can consist of more than one unit, but the number is variable.

The function of antibodies

The reason why these classes and subclasses of antibody are so important is that they activate different effector functions. For example, IgM is efficient at activating complement, leading to the lysis of a cell to which it is attached. IgE molecules bind to mast cells and, when they interact with antigen, the mast cell degranulates, resulting in release of histamine and other substances, which in turn initiates a local and acute inflammatory reaction around the site of residence of the mast cell. IgA is predominantly induced by chronic exposure at *mucosal surfaces*. These are internal surfaces of the body to which access is possible without disruption of the skin. Examples are the lungs, the gastrointestinal tract, and the urogenital systems.

IgA does not activate known effector functions such as acute inflammation, complement, or phagocytosis. The IgG subclasses have different and diverse activities, such as facilitating phagocytosis and initiating the activation of complement to lyse cells to which the antibody is attached. Moreover, these different classes and subclasses of antibody are optimally produced under different conditions. We shall see how significant such differential production is in understanding different clinical conditions.

Chapter 2

The immune system's pertinence to medicine

Five areas of medicine are related to the immune system. I now list them and give an indication of why I think an understanding of the basis of immunological tolerance and immune class regulation is so medically important. I argue that such an understanding will likely provide a basis for strategies of prevention and intervention in clinical situations in each of these areas.

Autoimmunity and transplantation

In the early 1900s, Paul Ehrlich raised antibodies against the red blood cells of one goat by injecting them into another goat. The antibodies so raised reacted with the red blood cells of the donor goat but not with the red blood cells of the recipient. This finding struck Ehrlich. It led him to suppose there must be a mechanism to prevent immune responses against self-antigens, leading to the attribute now known as self-nonself discrimination. He surmised that, if this were not the case, the immune system would be a greater threat to an individual than all external threats.[3]

Ehrlich's views were shown to be prescient in the 1940s, when it first became clear that some diseases were due to an immune attack against parts of the patient's own body. Such damaging immune reactivity is referred to as *autoimmunity*. One of the earliest autoimmune diseases to be recognized was ***autoimmune hemolytic anemia***. In this disease, antibodies are produced against the patient's own red blood cells, resulting in their destruction. The patient thus becomes anemic. Other common autoimmune diseases are ***autoimmune diabetes*** and ***multiple sclerosis***.

A second area of medicine related to the immune system is transplantation. In both autoimmunity and transplantation, we will have a much better chance to prevent autoimmunity and immune rejection of a graft, or intervene to ameliorate such immune responses once they have occurred, if we understand how antigen acts differently to ablate or stimulate immune responses; in other words, if we understand the basis of self-nonself discrimination. I will later argue that this intuitive judgment is supported by research in the field.

The three other major areas of medicine that are immune system-related are allergies, cancer, and infectious diseases. We start with infectious diseases.

Infectious diseases

We have already seen that immunologists have had remarkable successes, in fact their greatest triumphs, through vaccination in preventing such diseases as small pox, diphtheria, and tetanus and, in the 1950s, polio. However, it has not yet been possible to vaccinate effectively against some other infectious diseases. Why might this be?

All vaccination that is standard practice in western society, for example against small pox and polio, leads to a secondary antibody response upon natural infection. Vaccination is effective because antibody can limit an infection of the causative pathogen or neutralize the toxin the pathogen produces. The antibody ***neutralizes*** the toxin by binding to it in a manner that prevents the toxin from binding to a cellular receptor. Such binding by the toxin to the cellular receptor would otherwise initiate the pathological and toxic process.

However, a number of infectious diseases cannot be prevented by the kind of vaccination that leads to copious antibody production upon natural infection. Two important and illustrative examples are AIDS and tuberculosis. We need to understand the reasons for failure if we are to overcome them.

First, some pathogens are not readily contained by antibody, as is the case with **Trypanosoma brucei**. This protozoan is an extracellular parasite that causes sleeping sickness in humans. Successive waves of parasitemia follow infection. It turns out that the parasite has a major surface protein against which antibody is produced. These surface molecules pack tightly on the parasite's external surface, like pine trees in a forest. The antibody produced against this surface protein is highly effective in killing the parasite, whose numbers consequently drop dramatically. However, the parasite has a genetic mechanism that allows a few of them to switch the major surface protein expressed to a new variant, so that the antibody produced against the first wave of parasites is no longer effective in killing the switched parasites, giving rise to a second wave of parasitemia. Antibodies are produced against this second wave, leading again to a dramatic drop in parasitemia. Some parasites arise expressing yet another major surface antigen. This cycle repeats itself, leading to many successive waves of parasitemia. This process of escaping from the immune system on the part of the parasite is referred to as **antigenic variation**. It occurs in a slightly different manner in pathogens responsible for different infectious diseases. One of these is **acquired immune deficiency syndrome**, or **AIDS**, caused by the **human immunodeficiency virus**, or **HIV**.

The case of HIV

Most individuals infected by HIV succumb to AIDS. However, a small fraction of individuals does not. A study carried out in Kenya among sex workers shows somewhere around 5% of these individuals do not develop AIDS. It turns out that these **healthy infected individuals** generate a cell-mediated response and produce little if any antibody.

Moreover, these sex workers are exposed through their work to multiple viral variants, often different clades of the virus, yet do not succumb. In contrast, infected individuals that produce substantial anti-HIV antibody display the progressive stages of AIDS. One might suspect from these observations that an exclusive cell-mediated response is protective, whereas a predominant antibody response is not.

This supposition is supported by further observations. It is known that the nature of the immune response following infection most often evolves with time. First, a cell-mediated response is generated and then, somewhat later, antibody production usually occurs at an increasing pace. This pattern is seen following exposure to many non-living antigens as well as living infectious agents, such exposure occurring by many different routes. Exposure of people to HIV usually occurs through sexual intercourse or, years ago, from the transfusion of contaminated blood.

Interestingly, HIV-infected individuals do not suffer serious symptoms when their immune response is in a predominant cell-mediated phase. A short time after infection and before anti-HIV antibody is readily detectable, individuals are relatively well. It is only after they **seroconvert,** or start producing substantial levels of antibody, that the successive stages of debilitating AIDS occur. The period after infection, when they do not suffer from serious symptoms, is called the **honeymoon period**. The existence of this period again shows that a predominant cell-mediated response is sufficiently protective that the individual has minimal symptoms.

Why might antibody not be protective? Antibody produced upon infection can neutralize and protect against the virus that causes the initial infection. Such neutralizing antibody binds to the virus, thereby blocking the virus's ability to bind to receptors on target cells, a step necessary for viral entry into target cells in which the virus replicates. The antibody/virus complex may be **opsonized** by a macrophage, that is taken into the interior of the cell, and then destroyed. However, HIV has a way of circumventing the effectiveness of **neutralizing antibody**.

The virus has mechanisms to rapidly produce diverse variants within the infected individual. If antibody is produced that is effective in neutralizing and so protecting against one viral variant, its companion variants, that are not neutralized, will be at an advantage and will preferentially replicate. In general, the production of substantial antibody responses and diminished cell-mediated responses correlate with disease rather than health.

On this basis, some individuals, including me, suggest that if vaccination is to be successful against HIV, it must guarantee a predominant and protective cell-mediated response upon natural infection. We shall later consider how this might be achieved.

Tuberculosis

Tuberculosis is still the cause of more deaths than any other infectious disease. Current vaccination strategies barely protect the vaccinated individual. I shall argue later that a vaccination strategy that guarantees, upon natural infection, a strong and predominant cell-mediated response to *M tuberculosis* will be effective in protecting against disease. How this might be achieved again requires an understanding of immune class regulation.

Allergies

The incidence of allergies depends upon geographical location and season, as those antigens that cause allergies are often more prevalent in some places, and during some seasons of the year, than in others. In a geographical area where allergy is prevalent to a particular antigen, a state associated with the production of IgE antibody to this antigen, it is found that the non-allergic individuals living in this area are also immune, but have relatively high levels of IgA and IgG_4 and relatively low levels of IgE antibodies to the antigen. If we properly understand how different classes and subclasses of antibody are differentially regulated, we might be able to modulate the response from one antibody mode to another. If we could modulate the nature of the antigen-specific antibody response evident in allergic individuals to the type present in non-allergic individuals, we would presumably have an effective way of treating allergies. In addition, an understanding of the mechanisms underlying immune class regulation may also allow us to vaccinate individuals in such a manner to predispose them to generate, on contact with the allergy-inducing antigen, the immune state seen in non-allergic individuals, rather than that seen in allergic individuals.

Cancer

Human cancers and experimental tumors are recognized by the immune system

In the early 1900s, Paul Ehrlich speculated that the clue to controlling cancer was to understand how the immune system naturally responds against it.[4] This speculation only made sense if Ehrlich envisaged that cancer and normal cells were chemically different, and that the immune system could, and would, immunologically respond against the cancer cells as foreign. The question of whether the immune system responds against naturally arising cancers, in the manner Ehrlich envisaged, dogged the field for the better part of a hundred years. However, evidence in the last twenty years has convinced most, including me, that human cancers and experimental tumors do stimulate immune responses. We shall later indicate the nature of this evidence. The question then arises, why are cancers not contained?

There are some grounds for questioning whether most studies in animal models of human cancer actually provide useful and valid information concerning naturally arising human cancers. Nevertheless, it is useful to consider what such studies in animals show, and we can then more readily assess their relevance to understanding what happens in human cancer.

Most studies in animal models are carried out with tumors that arose in inbred strains of animals. Tumor lines are established from the initial tumor. These lines grow in vitro and are employed as a source of tumor cells. Studies often involve injecting a sufficient number of tumor cells into a naïve animal of the same strain as that in which the tumor arose, so that the tumor grows progressively. This means, if the animal is not sacrificed at a certain point to avoid unnecessary pain, it would die. We refer to an injection of this number of tumor cells as a lethal challenge.

Researchers have explored how they might prevent this progressive tumor growth. For example, a long established means of vaccinating against a normally lethal challenge of tumor cells was discovered in the 1950s, and seems to be widely applicable in most but not all tumor systems.

Researchers usually inject the tumor cells at a site close to the surface of the body so that they can readily estimate tumor size. Researchers found that if they gave a normally lethal challenge to a naïve mouse, for example, and then surgically removed or excised the tumor when it has a noticeable but small size, typically about nine days after injection, the mouse would remain healthy. Challenge of such a **tumor-excised** mouse with what would normally be a lethal challenge, a month or two after excision, resulted not in progressive tumor growth but in resistance to the tumor. The immune system of an excised mouse had been immunologically primed to resist the tumor. This procedure is referred to as *excision priming*.

The discovery of a means of establishing immune-mediated resistance to a tumor led investigators to ascertain whether such resistance was due to a cell-mediated or a humoral response. It was found in nearly every case that resistance could be transferred to a naïve mouse by giving it cells from a syngeneic and resistant mouse but not by transferring serum. Cell-mediated immunity is protective. This protective cell-mediated response has been correlated in diverse situations with the generation of what are referred to as *cytotoxic T lymphocytes* or **CTL**. These CTL have antigen specific receptors and can bind to and subsequently lyse the target cancer cells, ie to their breaking apart.

Moreover, tumor progression in animal systems was noted to be often associated with the production of tumor-specific antibody.

These observations of the 1950s and 1960s led to a widely held view that animal tumors and human cancers can be contained by a cell-mediated, but not by a humoral, response. I will argue later for the view that most animal tumors grow progressively because the cell-mediated response is too weak to kill the tumor cells at a greater rate than they are being generated by multiplication of the tumor cells. The tumor thus grows progressively under these circumstances.

The weakness and insufficiency of the cell-mediated response most likely occurs for one of two main reasons. Either, the tumor is of such a nature that only an intrinsically weak immune response can be generated. Alternatively, the tumor can induce more vigorous responses, but a humoral response, associated with partial inhibition of the protective, cell-mediated response, occurs. I will argue later for the pertinence of this latter view for understanding why immunity sometimes fails to contain animal tumors.

The immune system and human cancer

The real question is whether there are similar reasons for why human cancers grow progressively, as those envisaged for animal tumors. There certainly are a number of reports in the literature where cancer containment is associated with a predominant cell-mediated response, and cancer progression with a predominant antibody response. However, it is not possible to carry out comparable experiments in humans as have been carried out in animals, for both ethical and practical reasons. We cannot do similarly incisive experiments and should therefore be cautious in drawing conclusions.

However, there are ways of testing the plausible idea that the cause of failure to contain animal tumors and human cancers is often similar. We can devise treatments of human cancer based on this premiss. If our framework is valid, the treatments should be efficacious and this would both support the framework and improve cancer treatment. We will return to this topic at some length once we have a better understanding of the foundational concepts as to how immune responses are regulated.

In conclusion, I suggest that understanding what is required to generate immune responses is pertinent to increasing intrinsically weak immune responses against cancer, and understanding immune class regulation is central to preventing or treating cancers that grow progressively due to the anti-cancer immune response having a significant and detrimental humoral component.

The take-home message

In summary, it appears that understanding how antigen can interact differently with cells of the immune system, to either ablate or generate an immune response, is likely important in devising strategies to prevent or treat various clinical situations. These include those associated with autoimmunity, transplantation of foreign grafts, and with prevention and treatment of some cancers that grow progressively due to the naturally-generated immune response being too weak to contain the cancer.

Understanding the basis of immune class regulation may provide a conceptual platform for preventing and treating allergies, and so allergic-dependent asthmas, for preventing and treating diverse infectious diseases best contained by cell-mediated immunity against the causative pathogen, and for preventing and treating some cancers. These prospects provide an exciting context for examining the basis of immunological self-nonself discrimination and immune class regulation.

Chapter 3

The Clonal Selection Theory
of Antibody Formation

Instructive and selective theories of antibody formation

We have seen how Landsteiner's studies in the first half of the 1900s gave rise to an appreciation of both the specificity of antibodies and the universality of the antibody response.[1] These two attributes clearly implied that the repertoire of different antibodies an individual can produce is vast. Ehrlich had also pointed out in the early twentieth century that there must be a means of ensuring that anti-self immune responses are not readily generated.[3]

From the early 1900s to the 1940s, these facts were recognized but little significant discussion took place as to how the diversity of anti-bodies could be realized or self-nonself discrimination be achieved. Then, in 1940, Linus Pauling, a famous chemist also interested in biology, proposed a solution to the question of how the great diversity of antibodies might be realized,[6] a solution I shall shortly describe. In 1949, Burnet and Fenner wrote a book entitled *The Production of Antibodies*,[2] which was something of a landmark. It was a major step in taking a broad look at the various features of immune responses, listing them, and then questioning whether these features could be explained in terms of the model that Pauling envisaged. It is often the case in speculative science that careful and subsequent deliberation allows the issues to be crystallized and so be more effectively considered. In the following paragraphs, I shall not follow the detailed historical path of describing exactly what Pauling proposed and the response by Burnet and Fenner, but rather attempt to distil the issues by using their proposals to illustrate general points.

There are two exclusive views as to how the diversity of antibodies is generated. Pauling's 1940 theory is an example of one type of approach.[6] Proponents envisaged that antigen is needed to make a complementary antibody, and such suggestions are referred to as instructive theories, as it is postulated that the antigen is required to instruct the immune system to make a complementary antibody. Pauling's proposal was based on the knowledge that antibodies are proteins made up of polypeptide chains. He proposed that these chains are highly flexible and fold up around an antigen molecule, and retain their shape once the nascent antibody and antigen molecule come apart. However, there were also other instructive theories proposed around this time. What characterizes them all is the proposal that antigen is needed to make the corresponding antibody.

Others envisaged an incompatible scenario. They proposed that diverse antibodies are somehow generated in the absence of antigen. Antigen binds to and so selects complementary antibodies, resulting in their further and usually massive production. These theories are incompatible because **instructive theories** propose that antigen is needed to make a complementary antibody, while **selective theories** propose that the antibody pre-exists its selection by antigen. If not present, it cannot be selected.

Burnet and Fenner pointed out in 1949 that there were some features of immune responses difficult to reconcile with instructive theories.[2] First, there is a lag period of at least several days after antigen impact before there is detectable antibody, after which antibody is produced in an escalating manner. How could Pauling's theory or other instructive mechanisms explain this, particularly the escalating tempo?

Second, substantial amounts of antibody can be produced against a non-replicating antigen for prolonged periods, a fact that is again difficult to explain using these theories. Lastly, what could be the basis of immunological memory as seen in secondary immune responses?

Burnet and Fenner made a first proposal as to how self-nonself discrimination might be achieved. They suggested that all self-antigens are chemically marked early in development, preventing their completion of a step required to generate immune responses. It was clear that in this case a foreign antigen, introduced to an animal early in its development, would also become chemically marked and so be regarded by the immune system as a self-antigen. The capacity to respond against it as a foreign antigen would have been ablated when the animal developed immunocompetence to respond to other foreign antigens. Medawar and Hasek and their colleagues verified Burnet and Fenner's prediction in the early 1950s. Burnet and Medawar shared the 1960 Nobel Prize in Physiology or Medicine for these contributions to our understanding of self-tolerance.

The impact of the birth of molecular biology

A number of different proposals as to the origin of antibody molecules were made in the 1950s, culminating in the Clonal Selection Theory as we now know it. Several of its critical predictions were successfully tested in the 1960s, and it had become the cornerstone of the field by the 1970s. No one seriously questions its tenets today. The formulation of this theory gave immunology an incredible vitality for at least a couple of decades.

There were influences on the formulation of this theory from other areas of biology, molecular biology in particular. It is useful to note in this regard that Watson and Crick's proposal for the structure of deoxyribonucleic acid, DNA, was published in 1953. A knowledge of this structure had one of the most profound impacts on biology of any discovery in any field of science.

It had become apparent in the 1940s that the genetic material, most generally DNA, has two characteristics that make its role in defining the nature of living organisms central. First, this molecule has the ability to replicate. This property mirrors what might be regarded as the cardinal attribute of life. Second, this genetic material, present in the nucleus of animal cells, somehow specifies in large measure the nature of the cell. We need to understand in broad terms how DNA, and the environment in which it exists, results in the realization of DNA's two characteristics of replication and of determining the nature of the cell to which it belongs. This understanding, as it was achieved, had a profound impact on biology in general and on immunology in particular. Its impact was critical in the formulation of the Clonal Selection Theory.

Watson and Crick's proposed structure of DNA was so exciting because it offered a compelling insight into how DNA replication could occur. The DNA molecule itself consists of two entwined chains. They have direction in the sense that the two ends of a given chain are dissimilar, so that one end can be usefully defined as the beginning and the other as the end. The two entwined chains of a DNA molecule run in opposite directions, and are themselves made from linking together four bases, abbreviated as A, T, C, and G, in a regular manner. Thus, the chemical nature of a DNA chain can be defined by the order of its bases, e.g. AACGCCTT. Watson and Crick's structure of DNA is based upon the rule that if one chain has A at a particular position, the entwined chain must have T at this position, and if one chain has C at a position, the entwined chain has G at this position. The sequence of one chain thus defines the sequence of its entwined partner. These rules of base pairing have a simple, structural basis, similar to why one piece of a jigsaw puzzle only fits another piece: only A fits with T, and C only with G. This base-pairing rule provides the underlying idea for Watson and Crick's proposal for how DNA replication (duplication) occurs. The two entwined chains come apart and each acts as a template to define its complement, so one double-stranded parent molecule gives rise to two double-stranded daughters. This mechanism can be appropriately seen as an instructive process.

The second fundamental feature of DNA is that it somehow defines the characteristics of the cell in which it exists. There are three fundamental ideas pertinent to understanding how this is accomplished. The first relates to the central importance of proteins in determining the characteristics of a cell. Many proteins function as enzymes, and a particular enzyme will facilitate a particular chemical reaction. Thus, the enzymes a cell has determine the molecules it can make and greatly influences the chemical composition of the cell. Moreover, many structural components of the cell are also made of proteins. These structural and enzymatic functions of proteins explain their central role in determining the properties of a cell, and these functions of proteins depend upon their three-dimensional architecture. A protein usually loses its function when this architecture is altered.

The second idea is that the DNA exerts is genetic function in determining the characteristics of a cell by defining the proteins the cell can produce. We have to know more about the chemistry of proteins to appreciate how the DNA of a cell can specify the proteins it makes.

A given protein consists of one or a few polypeptide chains. These have a direction, in the sense that the two ends of the chain are different. Sanger first showed that a given polypeptide chain is made from linking together different amino acids in a definite and regular manner. Twenty different amino acids are found in polypeptide chains, each having a different and characteristic side chain. Sanger showed that each of the two polypeptide chains of insulin has a unique amino acid sequence. Let us for convenience refer to the twenty amino acids as "a" through "t". The chemical nature of a polypeptide chain is defined by specifying the order in which its amino acids are strung together, for example, d.t.m.a.a.q.p, and so on.

Polypeptide chains and DNA strands are thus both linear-type molecules made up from four, in the case of DNA, and of twenty, in the case of polypeptide chains, chemical units or building blocks.

Moreover, classical genetics and biochemistry had led to the idea that a segment of DNA, called a gene or cistron, codes for one polypeptide chain. What was meant by the term "code" was obscure. The unravelling of the process that constitutes coding is one of the major triumphs of molecular biology. In brief, it was found that three bases, a triplet or codon, of a DNA chain coded for one amino acid, and so a stretch of DNA consisting of three hundred bases can code for a polypeptide chain consisting of one hundred amino acids.

The third central idea has already been tacitly assumed in the foregoing account, but must be made explicit. The section of DNA that constitutes a gene specifies only the order, or the sequence, of the different amino acids of the polypeptide chain that the gene encodes. This idea is known as the Sequence Hypothesis. Central to it is the ancillary idea that proteins, as they are made in the cell, automatically acquire their particular three-dimensional shape or architecture that is central to their structural and enzymatic function.

The Clonal Selection Theory is born

Jerne's contribution

During his graduate studies, Niels Jerne was examining the production of antibody to a particular antigen. The assay for antibody that he employed was extremely sensitive, in the sense that it could detect exceptionally minute amounts of antibody. Jerne found that naïve rabbits, that neither he nor anyone else had immunized, had detectable levels of antibody to the antigen in their blood. This finding appeared to be consistent with selective theories, and seemed to mitigate against instructive ideas.

Jerne's finding led him to propose, in 1955, the first modern selective theory[7] since Ehrlich's in 1901.[8] He made this theory in the context of Burnet and Fenner's suggestion that the presence of self antigens, early in development, led to an ablation of the ability of the animal as an adult to respond against self antigens, and Medawar and Hasek's reports, two years previously, supporting this suggestion. Jerne made his theory at a time when it was realized that an individual can make a much larger number of chemically distinct antibodies than was envisaged by Ehrlich.

Jerne's theory had three major components. (i) A diverse array of antibodies is made early in development, and this array is represented by the presence of these antibodies in the blood. (ii) Those antibodies, specific for self-antigens present at this early time, combine with the self-antigens and are removed. These two processes leave a diverse population of antibodies in the blood of a developing person or animal with the potential to bind to foreign but not to self-antigens. (iii) When a foreign antigen impinges upon the body of an immunologically mature individual, the antigen binds complementary antibodies. The resulting antigen/antibody complexes are taken up by a phagocytic cell, whereby the antibodies direct the cell to produce more antibody molecules identical to themselves.

Jerne's was indeed a selective theory, in the sense that antigen selects pre-existing antibody molecules for replication. However, how antibody could replicate itself, or a protein provide instructions for the synthesis of a protein identical to itself, was obscure. This aspect of his hypothesis had the aroma of instructive theories. Jerne had not absorbed the tentative thoughts, later shown to be insights, of the instigators of what came to be called molecular biology. Only nucleic acids, namely DNA and RNA, have the ability to replicate and specify the nature of polypeptide chains. Jerne's third step envisaged that proteins themselves could have these abilities.

Burnet's contribution

Burnet no doubt contributed considerably to the ideas that came together to form the coherent framework now known as the Clonal Selection Theory. However, understanding what he wrote is often not easy, and his writing can be obscure when one compares his expositions to those of his contemporaries. It is my feeling that Burnet was perhaps not as analytical as others, but imaginatively faced big questions without intimidation. He championed both incorrect as well as novel and important ideas, and demanded less than others that published ideas are clear or consistent with a plausible framework. In the 1940s, for example, in discussing possibilities within an instructive framework, he thought that somehow the instructions might affect a cell so that it and its descendants could produce antibody complementary to the antigen. This instructive theory was again obscure from a mechanistic point of view, and clearly different from Pauling's flexible, polypeptide chain theory. Nevertheless, this idea was a first attempt to understand the basis of immunological memory, and it resurfaced as a central feature of **Clonal Selection Theory**. The idea is that different cells can have different, inherited potentials, so that these potentials are handed down from one cell to their progeny.

Talmage's contribution

David Talmage made a considerable contribution at this juncture, critically modifying Jerne's proposal. One gets the impression from his writings that he was much more careful, sparing, and considered in what he proposed than Burnet. Talmage appreciated the deeper issues being faced by those establishing the foundations of molecular biology. Proteins do not provide the instructions for their own replication, as Jerne's hypothesis demanded. Talmage proposed in 1957 that "one of the multiplying units of the antibody response is the cell itself...only those cells are selected for multiplication, whose synthesized product has affinity for the antigen injected. This would [require] a different species of cell for each species of protein produced..."[9] This statement can be regarded as constituting the critical step. One cell has the capacity to produce antibody of only one specificity. Therefore different cells are needed to make antibodies of different specificity. Talmage also included the idea that exposure to an antigen early in life would ablate the corresponding cells. The ability to produce antibody complementary to these antigens would be lost. He provided in this manner a

potential explanation for Hasek's and Medawar's observations on tolerance and for how self-nonself discrimination might be achieved.

A feature of Talmage's proposal is worth stressing. Given the great
specificity of antibodies and the universality of the immune response—
in other words, its ability to mount an antibody response against virtually all foreign invaders—the cells with antibody receptors specific for a
particular antigen, before this antigen first impinges upon the immune
system, must be very scarce indeed. Talmage envisaged that the cells
producing antibody arose through multiplication of precursor cells.

This idea was indirectly supported by two observations. There is a
lag period of several days following immunization before antibody
can be detected, and then the amount of antibody rises quickly. This
lag phase was naturally explained by the period during which the
stimulated but initially scarce precursor cells divided. In addition, it
was found that the ability of an animal to produce antibody against an
antigen was highly sensitive to radiation of the animal. Cells exposed to
radiation and forced into cell division usually die due to damage of the
few essential DNA molecules each cell contains. These observations are
readily understood if the scarce antibody precursor cells, specific for a
particular antigen, are triggered by this antigen to multiply during the
course of an antibody response, before they can give rise to antibody-
producing cells.

Talmage's proposal was clear, short, and succinct. It seems to me to
be the first published and coherent account of the essential elements of
the Clonal Selection Theory.

Lederberg's contribution

In 1959, two years after Talmage's proposal, Lederberg, the bacterial geneticist, published an article in *Science* entitled "Genes and
Antibodies".[10] This provides the clearest, most succinct, and extensive
exposition of the Clonal Selection Theory in my view. Burnet also
published his book, *The Clonal Selection Theory of Acquired Immunity*,[11]
in 1959. Lederberg had visited Burnet in Australia in 1958, and

corresponded with him after he became interested in the problems posed by the immune system.

Who formulated the Clonal Selection Theory?

My conjectures, stimulated by words on the printed page rather than by hear-say, is that Lederberg's contributions to the later phases of the formulation of the theory might be underappreciated, as are Ehrlich's and Talmage's contributions in the earlier phase. I know little of the personalities involved, but both Talmage and Lederberg's writings strike me as exceptional in their clarity. This feeling comes from studying many of the original articles and from biographical material available on the major participants. I understand Burnet thought an award of a second Nobel Prize to himself for the Clonal Selection Theory would have been in order. My comments above have been partly colored by his expression of this thought.

Several individuals no doubt made critical contributions to the theory, and it is probably unfair and an oversimplification to give primary credit to a single person. I am not a historian and have not studied the evidence to be confident to attribute credit justly, even if this is possible. However, what is most important is to appreciate the considerable degree to which many detailed subsequent observations fit compellingly into the theory's framework, and to appreciate that this theory provides us with the setting to achieve greater understanding.

Evidence for the Clonal Selection Theory

Though a tremendous amount of detailed evidence supports the Clonal Selection Theory, it is not necessary, in discussing foundational concepts of the field, to describe all of them. I just outline here the kind of experiment that supports major expectations based on the theory.

The theory accounts for the sensitivity of the generation of an immune response to radiation when an animal is irradiated within a day or two of immunization. This makes sense if, as already discussed, scarce antibody precursor cells need to divide extensively upon activation before their progeny differentiate and secrete antibody.

The sensitivity of antibody precursor cells to radiation was exploited by immunologists to good effect. They envisaged that irradiated mice, unable to make an antibody response, would make a strong antibody response upon immunization with antigen if they were reconstituted with unirradiated antibody precursor cells. They showed that the ability of irradiated mice to produce antibody could be restored by intravenously injecting them with spleen cells from syngeneic mice. This restoration implies that spleen cells contain antibody precursor cells.

Further observations strongly support this inference. About a third of mouse spleen cells bear on their surface antibody receptors, as envisaged of antibody precursor cells. Moreover, removal of these cells from the spleen cell population abrogates its ability to reconstitute a lethally irradiated animal to produce antibody. This finding is consistent with the idea that antibody precursor cells bear antibody receptors on their surface.

Moreover, it is found that scarce cells in the spleen will bind a foreign antigen. One example, used in many classical studies examining the generation in mice of antibody responses, is the antigen sheep red blood cells (SRBC). It is found that there are scarce cells among mouse spleen cells that bind to SRBC via antibody receptors on their surface. It is possible to see such cells under the microscope, as they bind so many red blood cells. Due to their appearance, they are called rosettes. It is also possible to selectively remove them from a spleen cell population. Such depleted spleen cells can no longer restore the response of irradiated mice to SRBC, but they can restore the response to other antigens, such as chicken red blood cells (CRBC). This particular experiment shows that different cells are responsible for generating antibody responses to SRBC and to CRBC. The cells required to generate antibody to SRBC have antibody receptors that bind to SRBC, and the same applies for CRBC. This type of analysis has been carried out in many different systems. It tests the central idea that antibody precursor cells have antibody receptors on their surface able to bind antigen. In addition, those cells able to bind the antigen are required for making an antibody response to this antigen but are not required for making antibody responses to unrelated antigens. Further evidence showed that, on interacting with antigen, these receptor-bearing cells multiply and their progeny secrete large amounts of antibody.

Testing Lederberg's further speculations

Lederberg made two further speculations in his 1959 paper that I believe to be unique, and which proved to be remarkably insightful. It took decades before these speculations were found to be correct. Lederberg's paper was written by someone clearly neither faint of heart nor of mind. As an aside, my comments here are colored by the fact that I read four obituaries for Lederberg when he died, but none mentioned his 1959 *Science* paper, "Genes and Antibodies", which I consider one of the most remarkably insightful and beautiful scientific papers I have ever read.

Cellular Model for Self-Tolerance

All those who contributed to the formulation of the Clonal Selection Theory, except Lederberg, followed Burnet in postulating that self-nonself discrimination was uniquely established during development. There were three important aspects to Lederberg's alternative proposal. First, he suggested that antibody precursor cells are continuously generated throughout an individual's life, and so self-nonself discrimination must be a continuous process. Autoimmunity can strike an individual in mid-life, a fact suggesting self-nonself discrimination is an on-going process.

Second, given this surmise, it would appear that the lack of immune responses to self antigens requires not only the presence of self antigens early in development, but also their continuous presence thereafter to maintain tolerance. This is easy to demonstrate in an experimental model of self-tolerance. For example, fertilized chicken eggs take twenty-one days to hatch. It is possible, at about day twelve of gestation, to drill a hole in the shell and inject foreign antigens into a vein of the developing chick. Researchers showed in this way that injecting turkey red blood cells (TRBC) into the developing chick resulted in the young adult chick being unable to make an antibody response to TRBC.

Turkey red blood cells do not replicate, and are cleared from the chick's body at a significant rate. In the absence of TRBC, new anti-TRBC antibody precursor cells are generated and survive. If the developing chick is only exposed at day twelve of gestation to TRBC, it thus regains its ability to respond to TRBC as an older adult. It is necessary to give the TRBC periodically to maintain their presence and so maintain the unresponsive state. This example illustrates the correctness of Lederberg's proposal that the continuous presence of self antigens is required to maintain the unresponsive state. We refer to his vision of what is required to achieve and maintain self tolerance as the Historical Postulate. This postulate proposes that unresponsiveness against self antigens requires their early presence in the development or history of the individual, and their continuous presence thereafter, see Figure 4.

Third, Lederberg proposed a precise model at the cellular level for how tolerance is achieved. It took some decades before its pertinence to the mechanism of self tolerance was demonstrated.

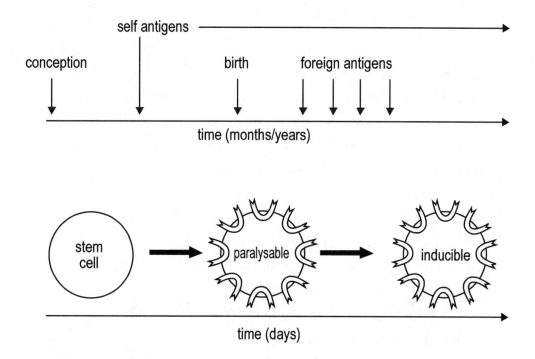

Figure 4. The Historical Postulate and Lederberg's proposal as to how self-nonself discrimination is achieved

Lederberg proposed that, when antibody receptor-bearing precursor cells first arise, they are paralyzed, die, or are inactivated if their receptors engage with antigen, as depicted in Figure 4. Thus, if their receptors interact with a self-antigen, the interacting cell would be silenced. However, if their receptors do not interact with antigen over a few days, an internal clock signals the cell to differentiate into a new state. This cell bears the same antibody receptors as the one from which it arose, but is now programmed to be activated when its receptors interact with antigen. Thus, it becomes what we have previously referred to as an antibody precursor cell. On interacting with antigen, the cell multiplies and its progeny differentiate into antibody producing cells.

Lederberg's model explained how anti-self cells could be ablated, and how a population of antibody precursor cells could be generated with antibody receptors that could only interact with foreign antigens.

Model for the origin of antibody genes

Lederberg's most remarkable and successful speculation, in my opinion, was directed at how the genes coding for the light and heavy chains of antibody molecules might be generated. There are two distinct aspects of his proposal. The first was the time frame within which he envisaged the genes to be generated. According to the common wisdom of the day, genes were generated through evolutionary time. According to this view, the genes I have received are some of those my parents had, unless slightly modified by the occasional mutation. Lederberg proposed something radically different. He proposed that we each individually generate the genes encoding light and heavy chains in special cells, and that once genes encoding light and heavy chains have been generated, these cells can give rise to an antibody bearing cell. In accord with this suggestion, he envisaged that what we receive from our parents are kits to assemble (Lederberg's word) genes from smaller units. I take his words to indicate, in broad terms, that he foresaw what we subsequently discovered to be the case.

About fifteen years later, Tonegawa provided the first evidence in favor of Lederberg's vision, and a new field was established. I will briefly describe these advances, but the small space I devote to them in no way reflects their importance to the field. The understanding of the mechanisms by which immunoglobulin genes are assembled is detailed and compelling, and I do not anticipate that what is now generally believed will be overturned by future progress. In this sense, the success of this large field of research means that I, for one, do not anticipate it to become a frontier of radical change. Therefore, it is sufficient to outline the broad conclusions made. However, I hope the brief sketch I provide of these advances is adequate for the reader to get a glimpse of these miracles of nature and to appreciate how remarkable our progress has been. Readers interested in a more detailed account of these advances may wish to consult an article *in Scientific American,* reference 12, written by Susumu Tonegawa.

The Generation of the Diversity (GOD) of antibody genes

We have seen that antibody molecules are made up of units containing two light and two heavy chains, and so consist of two identical halves. Moreover, the antigen-binding site of the antibody molecule is made up from a combination of the variable domains of the light and heavy chains. We can construct, if we have a thousand different light and a thousand different heavy chains, a thousand times a thousand different antibody molecules, or a million. Perhaps only 10% of these combinations fit well together, and so only a hundred thousand different, functional antibody molecules might be generated. This example illustrates the power of combinatorial systems to generate vast numbers of different antibody molecules. From two thousand genes, nature can produce a hundred thousand different antibody molecules! Nature has used this combinatorial technique time and again to generate an incredible diversity of genes coding for light and heavy chain variable regions employing a limited amount of DNA. What's more, we have the potential to generate many more light and heavy chains than a thousand, a number I chose only to illustrate the power of combinatorial mechanisms to generate large numbers of antibody molecules.

The children's toy Lego of four decades ago was different from the modern version. An old Lego set was designed to build an unbelievable variety of constructions, rather than to construct a particular object, such as a space shuttle. With a few hundred pieces, many different constructions could be made. Antibody genes are constructed according to similar principles.

I illustrate this process by explaining the rudimentary steps in assembling a gene coding for a heavy chain. These processes occur in what is called a pre-antibody precursor cell because the processes, when successful, result in the birth of an antibody precursor cell.

We inherit from our parents a set of V_h DNA segments, about 150 or so different variants, 12 different D_h segments, and 5 J_h segments. A complete V_h gene is assembled by joining together one of the 150 V_h segments to one of the 12 D_h segments, and this $V_h D_h$ unit is then joined to one J_h segment. We can make 150 x 12 x 5 different combinations, totaling 9000. However, there are further mechanisms to produce more variants. For example, an enzyme inserts a random DNA base or two between these segments as they are joined. This assembly process can in fact produce more than a million different heavy chain genes. Note that there are several random occurrences during this assembly process, and this randomness has two consequences. Different genes are assembled in different cells, and often an assembled gene does not code for a functional heavy chain.

The pre-antibody precursor cell has a means of assessing whether the heavy chain is potentially functional. If it is, it tries to assemble a functional light chain gene. If this too is successful, the cell stops trying to assemble further light chain genes. If all the attempts are unsuccessful, the cell dies.

Because the cell has a mechanism for assessing the potential quality of its assembled heavy and light chain genes, and stops when the cell appears to have been successful, one cell will only be able to make antibody of one specificity. This explains at the molecular level how the postulates of the Clonal Selection Theory are realized. It is truly an amazing story. In practice, the various pre-antibody precursor cells can give rise to millions of diverse antibody precursor cells that have genes coding for different light and different heavy chains. When I read what Lederberg published in 1959, I guess that he visualized an assembly process like this.

Chapter 4

The Clonal Selection Theory becomes the framework for further investigations

New areas of investigation

The Clonal Selection Theory reached its mature formulation at the end of the 1950s. Many immunologists in the 1960s and 1970s naturally assessed its plausibility and put their minds to testing its predictions. However, four areas of investigation arose in the 1960s and the 1970s whose aims were not to directly test the theory. These areas later became the subject of intense research. We sketch the early observations made in these four areas to set the scene for a later and fuller description of the subjects of these investigations.

(i) Generating unresponsive states in new borns

We have seen that in the early 1950s, Hasek and Medawar and their colleagues provided observations consistent with Burnet and Fenner's proposal for how tolerance is established. Further experiments, with the aim of extending these observations, were carried out primarily in five species of animals, namely chickens, guinea pigs, mice, rabbits, and rats. These experiments were directed at gaining an insight into the mechanisms resulting in the ablation of anti-self immune responses. In all these species, with the exception of chickens, who develop in eggs, it was difficult to deliver foreign antigens to the developing fetus. Immunologists therefore examined whether it might be possible to induce tolerance in newborn animals. They found they could achieve this under some conditions. It is important to note that most of these studies examined only the antibody response. One example illustrates this kind of experiment.

Rabbits were injected with a substantial dose of the foreign protein **bovine serum albumin**, or **BSA**, which is highly prevalent in cow serum and so is easily purified and available. It was found that normal rabbits readily produce antibody when immunized with BSA at three months of age. However, rabbits exposed to a massive dose of BSA on their first day of life were unable to do so. Immunologists generally thought that these unresponsive states were appropriate models for the unresponsive state that naturally exists against self-antigens. Thus, these experimental systems were developed to investigate the nature of self-tolerance.

(ii) Requirements to generate an antibody response

Another study with BSA exemplifies many similar approaches. In this case, the antigen was not given to neonatal animals, but rather to older but still young animals that were immunocompetent, that is able to generate immune responses. Young guinea pigs produce antibody when immunized with BSA bought from a commercial source. However, a careful analysis of this commercial BSA showed it to be heterogeneous. The BSA molecule has a molecular weight of about 68,000, but the commercial preparation contains material of higher molecular weight. This heavier material consists primarily of BSA molecules aggregated together. When this aggregated material is removed, the preparation is referred to as **deaggregated BSA**. It was found that deaggregated BSA is not immunogenic. Moreover, if a guinea pig is exposed to a sufficient dose of deaggregated BSA, it will not produce antibody to a challenge of commercial BSA given one month later. The deaggregated BSA is said to be **paralytic**, or **tolerogenic**, whereas the commercial BSA is said to be **immunogenic**, capable of generating an immune response.

These observations are reflective of similar ones made with different antigens in different species. The general form of such experiments is outlined in Figure 5. It became accepted, following such studies, that the larger the foreign molecule is, the more immunogenic it tends to be, and that immunogenicity correlated with the degree of aggregation of the antigen. One further and particular example illustrates these conclusions on immunogenicity and became important for further analysis. It was found that some strains of guinea pigs make a brisk antibody response when challenged with the antigen *poly- L-lysine*, or *PLL*, whereas other stains do not. These two strains are respectively referred to as responders and non-responders. It was found that if PLL was conjugated to BSA, itself immunogenic in guinea pigs, the non-responders could now produce antibody to PLL on challenge with the conjugate. Thus, immune serum from a non-responder, previously challenged with PLL-BSA, would precipitate with the antigen PLL.

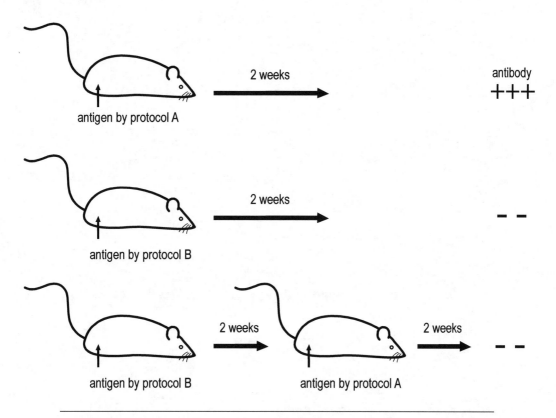

Figure 5. Protocol for demonstrating unresponsiveness in immunocompetent mice or other animals.

(iii) Humoral immune deviation

The third area explored in the 1960s was related to the existence of different classes of immunity, usually assessed in the form of DTH and IgG antibody. There were two major findings. Different conditions of immunization were found to favor the generation of DTH and the production of antibody. Indeed, Salvin had delineated the importance of the variables of antigen dose and time after immunization in a study carried out in the 1950s. He showed that a cell-mediated DTH response is usually generated first, and that this cell-mediated response declines as antibody is produced, see Figure 6. Increasing the dose of the antigen challenge increases the tempo of the response such that a DTH phase is often barely detectable if a sufficiently high dose of antigen is given. Conversely, a DTH response is uniquely but less rapidly generated if the dose of antigen is lowered sufficiently. Salvin's observations reflect, as previously noted, that there is a tendency for the generation of strong cell-mediated and antibody responses to be exclusive.

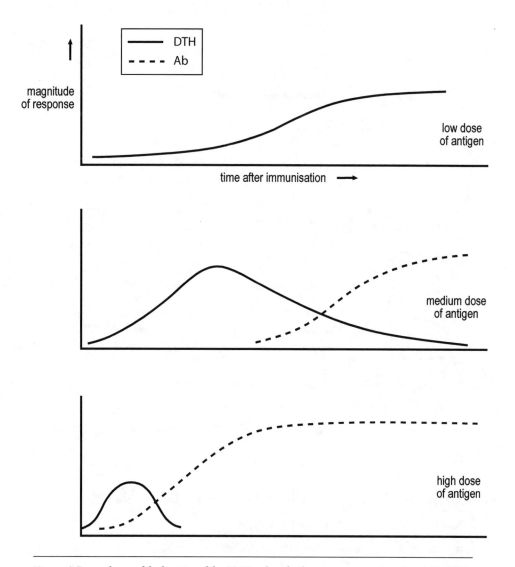

Figure 6. Dependency of the kinetics of the DTH and antibody response on antigen dose, after Salvin

Asherson and Stone first reported on what is presumably a related phenomenon in the mid-1960s. Their findings are summarized in Figure 7. Conditions were found under which antigen can induce in naïve animals a DTH response, and other conditions where the same antigen stimulates the production of antibody. They further found that animals immunized to produce antibody were no longer able to make DTH responses.

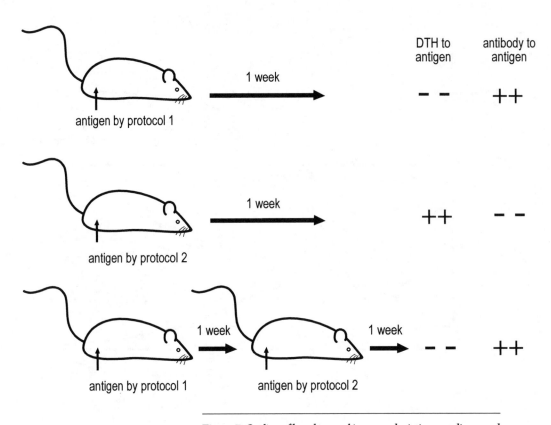

Figure 7. Outline of how humoral immune deviation was discovered

They called this phenomenon ***immune deviation***, though I call it ***humoral immune deviation*** to include the direction of the deviation. It appeared that the immune response to this antigen was locked into a humoral, antibody mode. We shall later see that immune responses can also be locked into a cell-mediated mode.

(iv) Lymphocytes are generated and immune responses are induced in different organs

Antibody precursor cells belong to a class of blood cell called a lymphocyte. Resting lymphocytes were originally characterized on morphological grounds. When activated, they increase in size dramatically, generating much machinery required to synthesize proteins and carry out other cellular functions. Resting lymphocytes, in contrast, consist of a nucleus with little extra-nuclear cytoplasm. In time, it was found that purified populations of blood lymphocytes could reconstitute irradiated animals so they could generate both primary cell-mediated and humoral responses. In addition, the reconstitution of irradiated animals with lymphocytes from primed animals able to generate secondary immune responses allows the reconstituted, irradiated mice to also generate secondary responses. Thus, lymphocytes appear to contain the precursor cells required for both primary and secondary immune responses. Lymphocytes are the immunocompetent cells.

Morphologically similar cells to blood lymphocytes are found in substantial numbers in two kinds of organ. For example, they are found in the spleen and lymph nodes. These are organs where antigens induce immune responses and are called secondary lymphoid organs. Lymphocytes are also found in the primary lymphoid organs that provide the environment that allows stem cells to give rise to functional lymphocytes. There are two major classes of lymphocyte. The B lymphocyte, more often referred to simply as a B cell, is generated in post-natal humans in the bone marrow. The T lymphocyte, more often referred to simply as a T cell, is generated in the thymus, an organ situated just under the top of the rib cage. T cells have diverse and important functions. The first to be recognized was as mediators of cell-mediated immunity, in the form of cells mediating DTH responses. Later studies showed that cytotoxic T lymphocytes (CTL) and DTH-mediating cells are different T cells.

Central tolerance and the need for peripheral tolerance

I break here from following the history of the field in my account. This break allows me to more readily and clearly describe our current understanding.

Lederberg's 1959 proposal as to how tolerance is achieved was not prominently considered in the mid to late 1960s, if at all. Some observations had been reported during the 1960s that were more easily understood in terms of a different framework from Lederberg's. In addition, there were no observations then that supported his proposed mechanism. Such support came later, in the 1980s. In more modern times, we have come to recognize that self-tolerance can be likened to an onion. There are layers of different mechanisms complementing each other in achieving this attribute.

The most important mechanism leading to tolerance is close to that first envisaged by Lederberg. This process occurs in primary lymphoid organs, where lymphocytes are generated, and is called ***central tolerance***. I first describe this mechanism and observations supporting it. This account then provides a context for appreciating its inadequacies in providing a sufficiently reliable mechanism of self-nonself discrimination, and so the need for ancillary mechanisms.

Central tolerance

Mature T cells have antigen-specific receptors, as B cells do. The ***T cell receptor (TcR)*** is made up of two polypeptide chains, the α and the β chains. Both these chains have two domains, the Vα and the Cα and the Vβ and Cβ domains. The genes, coding for these chains, are assembled by a process similar to that by which light and heavy chain Ig genes are assembled. A stem cell migrates to the thymus and, in this environment, differentiates into a pre-T cell. The processes involved in the assembly of α and β TcR genes is initiated in the pre-T cell. In time, some of these attempts result in cells expressing functional α and β chains, and so in the generation of T cells.

If these newly generated T cells have receptors that can interact with their antigen, present in the thymus, they are inactivated by a Lederberg-type mechanism. The antigens present in the thymus are predominantly self-antigens. The developing T cell has an internal clock; if it does not interact with antigen over a period of a few days, it differentiates into a mature T cell that, on interacting with antigen, can be activated (Figure 4). This mature T cell emigrates out of the thymus and circulates around the body, residing in and passing through the spleen and lymph nodes. B cells are generated by a similar process.

This scheme is consistent with Lederberg's proposed mechanism. However, we have come to realize there are three respects in which Lederberg's proposal is not the whole story, perhaps only 95% of it!

The limitations of central tolerance

First, if all self antigens were sufficiently expressed in the thymus, it is difficult to understand how autoimmune T cells could exist and emigrate into the peripheral circulation and then become activated. When immunologists examine autoreactive T cells, present in autoimmune disease, they usually find they are specific for self-antigens predominantly expressed in specialized cells present in organs. Insulin is such an example. It is primarily made in the specialized β islet cells of the pancreas. It is significant that antibody and T cells specific for insulin are found in individuals with autoimmune diabetes. This must mean insulin is not present in the thymus at a sufficient level in all individuals to ablate all T cells specific for it. This ablation process, when it occurs, is referred to as central tolerance. Self-antigens not present at a sufficient level in the thymus to cause reliable central tolerance are called *peripheral self-antigens*. I employ insulin as a peripheral self-antigen to illustrate the ways in which we now appreciate Lederberg's mechanism is an incomplete description of T cell tolerance.

Insulin can be detected in the thymus, but the level is clearly insufficient in some cases to ablate all T cells specific for it. Most interestingly, the incidence of autoimmune diabetes is controlled by many genetic loci. Alternative genes at one of these loci determine the level of insulin expression in the thymus. A higher level of expression corresponds to decreased susceptibility, presumably because a higher level results in more complete central tolerance.

The failure of central tolerance to reliably generate unresponsiveness to peripheral self-antigens allows us to understand the physiological need for, and the role of, mechanisms of *peripheral tolerance*. Thus, we shall see that T cells specific for insulin that emigrate from the thymus can interact in different ways with insulin, one way leading to their activation, the other leading to their inactivation. This framework is different from that proposed by Lederberg, according to which the mature T cells are unequivocally committed to be activated upon interacting with antigen.

Peripheral lymphocytes can be activated or inactivated by antigen

The idea that a mature lymphocyte can interact with antigen in two ways, one leading to its activation and the other to its inactivation, was the working hypothesis that most immunologists employed in the mid-1960s to mid-1970s. We have seen that different ways of administering antigen to immunocompetent animals can result in either an unresponsive state or the generation of an antibody response, as depicted in Figure 5. For example, guinea pigs given deaggregated BSA become unresponsive to immunization with commercial BSA.

Immunologists envisaged that these competitive processes at the level of the system reflect competition at the level of the individual lymphocyte. Thus, in the absence of any evidence in the 1960s for central tolerance, immunologists imagined that self-nonself discrimination was realized solely via a mechanism that we now identify as peripheral tolerance. This caused an awareness in the late 1960s of a major question: How does antigen interact differently with lymphocytes to result in their activation and in their inactivation?

Carrier effects

An extensive and careful reading of the literature up to the middle of 1967 leads me to suggest that most immunologists envisaged that antigen interacted in some way with an antibody precursor cell, resulting in its activation, in other words, its multiplication and the differentiation of its progeny to become antibody-producing cells. It is interesting to note that exactly what was involved in these different processes was for the most part not clearly and explicitly discussed. Implicitly, most who addressed this question envisaged that the binding of antigen to the antibody receptors of the precursor cell resulted in its activation. We call this the Simple Antigen Binding Model for the activation of the antibody precursor cell. For clarity, I have to comment that there was surprisingly little discussion at this time as to what was required for antigen to inactivate the antibody precursor cell. I stress this point because it may be as surprising to the reader as it is to me. Sometimes questions that in hindsight seem central are not clearly formulated and different possibilities are not discussed.

A number of observations accumulated in the literature throughout the 1960s that were difficult to reconcile with the Simple Antigen Binding Model. There were four areas of investigation where such irreconcilable observations were made. It was recognized that the phenomena described might reflect a common circumstance, and the puzzling aspect of these observations came to be referred to as *carrier effects,* for reasons that will soon become apparent.

(i) The requirements to generate secondary, anti-hapten antibody responses

A most interesting series of experiments was carried out by Mitchison. He examined the conditions required to generate secondary antibody responses to haptens. He immunized some mice with a hapten, h, coupled to a carrier, Q, i.e. with hQ. Such primed mice would naturally make a secondary antibody response against the hapten on challenge with hQ. The anti-h antibody produced in a primary response would of course react with hQ, but it also reacted well with the hapten coupled to a different protein, R, that did not crossreact with Q. In other words, the anti-hQ antibody reacted with hR, but not with R. When Mitchison primed mice with hQ and challenged with hR, he found the antibody response to h was much more like a primary rather than a secondary response. However, as the anti-h antibody produced on immunization with hQ reacted well with hR, this finding was paradoxical in terms of the Simple Antigen Binding Model for the activation of antibody precursor cells.

A provisional idea to explain these observations was that the receptor on the antibody precursor cell recognized the hapten as well as part of the adjacent carrier to which the hapten is attached. This and related phenomena were therefore referred to as carrier effects. However, this idea really violated the Clonal Selection Theory as it implied that the specificity of the receptor was different from that of antibody. As we shall see, this idea did not stand the test of time.

(ii) Breaking the unresponsive state

Weigle reported some highly intriguing observations in 1961. Rabbits can be made unresponsive to BSA by exposing them to a substantial dose of the antigen on their first day of life. We have seen that such rabbits at three months of age do not produce antibody on challenge with BSA. However, Weigle found a situation where he could immunize these rabbits to produce anti-BSA antibody.

Antibodies are highly specific. However, some antigens share structural features, and in this case antibodies to one antigen may also bind to another structurally related antigen. For example, immunization of a naïve rabbit with **human serum albumin (HSA)** results in the production of anti-HSA antibody, and about 10% of this antibody also binds to BSA. The two antigens, BSA and HSA, are said to **crossreact**. The antibodies that bind to both BSA and HSA are called **crossreactive antibodies**. We have seen that protection against smallpox virus is achieved by immunizing with cowpox virus. This protection occurs because the two viruses crossreact.

Weigle had found that rabbits, exposed to BSA on the day of birth, would not produce any anti-BSA antibody when challenged with BSA at three months of age, but small amounts of anti-BSA antibody are produced on challenge with HSA, see Figure 8. It turns out that this early experiment mirrored several reported later. I argue later that this experiment represents a most important situation under which antibodies to some peripheral self-antigens can be generated. We refer to these situations by saying that immunization with an antigen that crossreacts with an antigen, against which an animal is unresponsive, can sometimes **break the unresponsive state**.

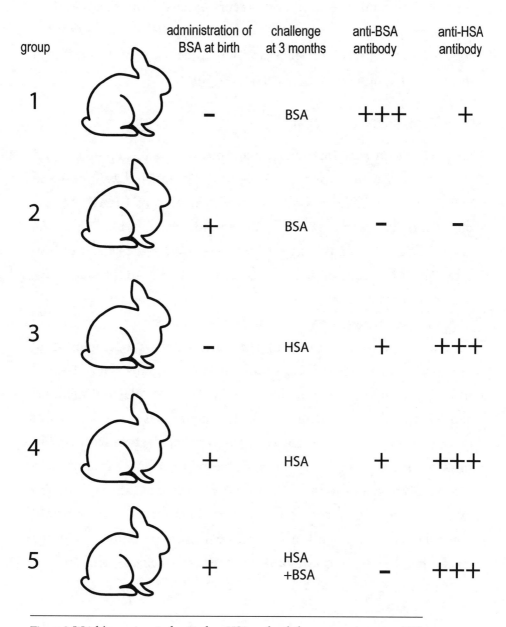

group		administration of BSA at birth	challenge at 3 months	anti-BSA antibody	anti-HSA antibody
1		−	BSA	+++	+
2		+	BSA	−	−
3		−	HSA	+	+++
4		+	HSA	+	+++
5		+	HSA +BSA	−	+++

Figure 8. Weigle's experiments showing how HSA can break the unresponsive state to BSA

According to the Clonal Selection Theory, the ability to generate anti-BSA antibody in BSA-unresponsive rabbits on challenge with HSA must mean that there are antibody precursor cells with receptors that bind to both BSA and to HSA.

In the context of the Simple Antigen Binding Model, it is paradoxical that these anti-BSA antibody precursor cells are not activated on challenge with BSA. These observations suggested to some that the specificity of the receptor on an antibody precursor cell was different from that of antibody. I argue later against this idea.

(iii) Macromolecular Haptens Exist

We have already seen that some guinea pigs are unable, upon immunization with PLL, to produce antibody to PLL, though they do respond to produce anti-PLL antibody upon a challenge of PLL-BSA. This observation is again paradoxical on the Simple Antigen Binding Model for the activation of antibody precursor cells. Why does PLL not activate anti-PLL precursor cells, which evidently exist, to produce anti-PLL antibody as anticipated?

A second example of a macromolecular hapten came from a study with rabbits, in which three related protein antigens were employed, all of which contain four protein subunits. Two contained four identical subunits, A_4 and B_4, and the other contained two A and two B subunits, and is referred to as A_2B_2. Rajewsky and colleagues found a strain of rabbits unable to produce antibody upon immunization with B_4, but they produced a vigorous antibody response upon immunization with A_4. Rabbits of this unresponsive strain produced strong antibody responses upon immunization with A_2B_2. The immune serum would precipitate with B_4. Thus, according to the simple antigen binding model for the activation of antibody precursor cells, B_4 should have been immunogenic.

(iv) Can macromolecular haptens inactivate antibody precursor cells?

Another finding exemplifies observations made in a few different experimental systems. We have just described the experimental system in which rabbits, unresponsive to B_4, can be immunized with A_2B_2 to produce anti-B antibody. It was known that animals made unresponsive to an antigen, BSA for example, remained unresponsive for several months, even if BSA is not further administered. In other words, it took several months to recover the ability to respond. It was probably in the context of this general finding that the following type of experiment was conducted with non-responder rabbits, unable to produce antibody on challenge with B_4. Rabbits were challenged with B_4 and challenged again a few weeks later with A_2B_2. These rabbits made a vigorous antibody response to B. It was concluded that B_4 was both non-immunogenic and non-tolerogenic, as it appeared that B_4 could not inactivate its corresponding antibody precursor cells. As indicated above, this was not an isolated finding. Immunologists generally agreed that non-immunogenic molecules were also non-tolerogenic.

An explanation of some carrier effects: cellular cooperation in the antibody response

Mitchison reported on some particularly insightful experiments in 1967. He was again delineating the conditions under which a secondary response to a hapten could be generated. We recall that priming with hQ resulted in a secondary anti-h antibody response on a secondary challenge with hQ, but not on challenge with hR. Mitchison showed that priming the same mouse with both hQ and R resulted in the mouse making a secondary anti-h antibody response on a subsequent challenge with hR. This led Mitchison to propose that, to successfully prime for a secondary antibody response to a hapten, it is necessary to prime separate cells to the hapten and others to the carrier.

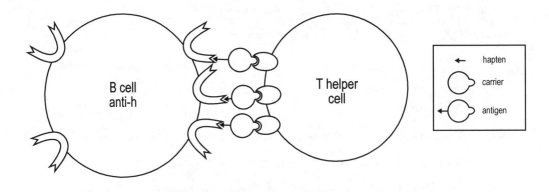

Figure 9. The Antigen Bridge Model of the B cell/T helper cell interaction

Rajewsky similarly concluded from his studies in B_4 unresponsive rabbits that the primary activation of an anti-B antibody precursor cell could be achieved on challenge with A_2B_2. However, anti-B antibody was not induced in the same rabbits when they were simultaneously immunized with both A_4 and B_4. These observations showed the anti-B antibody precursor cell could only be helped by A-specific "helper cells" if the A and B protein subunits were physically linked. Mitchison carried out similar experiments in his system, analyzing the requirements to generate secondary anti-hapten antibody responses. These observations naturally led to an Antigen Bridge Model, in which the activation of an antibody precursor cell is facilitated by a helper cell when they both bind to the same antigen molecules, see Figure 9.

Earlier observations on cell cooperation in the antibody response

These reports in 1967 had been preceded by another in 1966 that was puzzling in terms of the Simple Antigen Binding Model of B cell activation. Jacques Miller had demonstrated an immunological role for the thymus some years earlier. He had shown that the surgical removal of the thymus from mice within twenty-four hours of their birth resulted in gross immunological deficiencies once they matured into adults. Such neonatally thymectomised mice could not generate antibody responses to most antigens, and were totally unable to generate cell-mediated responses.

Irradiation can be lethal, as many of the white cells in the blood turn over, and so are continuously generated from dividing source cells. Irradiation kills the dividing source cells and results in a great deficiency of white and red blood cells needed to survive. Lethally irradiated mice can be saved by giving them unirradiated **stem cells**, which multiply and give rise to diverse white and red blood cells. Stem cells are found among the population of cells present in bone marrow. Henry Claman and colleagues tried to find out what the role was of thymocytes, the bulk of the cells in the thymus, in generating antibody responses. They tried to reconstitute lethally irradiated mice with thymocytes, or with bone marrow cells, or with a combination of both, and challenged all of them with the antigen SRBC. They found that only mice given both thymocytes and bone marrow cells produced high levels of antibody to SRBC. This experiment, reported in 1966, was the first to be interpreted as showing that more than one cell type is required to generate antibody responses.

Mapping the B cell and T cell interaction onto the interaction between hapten-specific antibody precursor cells and carrier specific helper cells

It was later found that some of the cells in the bone marrow express antibody receptors on their surface, and that the antibody-producing cells were descendants of the bone marrow population. The majority of the cells in the thymus have on their surface an antigen called the Thy-1 antigen. Separate cells bearing antibody receptors and the Thy-1 antigen are found among spleen cells, which is why these cells can be used to reconstitute lethally irradiated mice to generate antibody responses.

Moreover, neonatal thymectomy results in a gross deficiency of Thy-1 bearing cells in the spleen and other secondary lymphoid organs. Thus,

Thy-1 cells are appropriately called thymus-dependent lymphocytes, or T cells. The premise that the induction of antibody requires an interaction of B cells and T cells explains why neonatally thymectomised mice are usually deficient in their ability to generate antibody responses.

A particularly enlightening series of experiments was carried out with the system Mitchison had developed to analyze the requirements to generate secondary antibody responses to a hapten. He showed that irradiated mice, if reconstituted with spleen cells from mice primed with hQ, could make a secondary response to the hapten on challenge with hQ, but not on challenge with hR. He further showed that irradiated mice reconstituted with spleen cells from mice primed with hQ, as well as spleen cells from other mice primed with R, would make a secondary antibody response to the hapten on challenge with hR. It seemed that the hQ primed spleen cells must be the source of the memory, hapten-specific antibody precursor cells, i.e. of the memory B cells. The cells primed to R presumably provided the memory cells that played the role of thymocytes in the primary response examined by Claman. Working in Mitchison's lab, Martin Raff investigated the role of the R primed spleen cells in supporting a secondary anti-h antibody response on challenge with hR. He could abrogate this supporting role if the Thy-1 bearing cells were removed from the R-primed spleen cell population, before these spleen cells were given to the irradiated animal. The R primed cells, responsible for helping the anti-hapten response on challenge with hR, were T cells!

These studies, some performed by Av Mitchison and his group in England, others by Henry Claman's group in the United States, and yet others by Jacques Miller's group in Australia, had a decisive impact on the field. It was recognized that both primary and secondary antibody responses were optimally generated through cellular cooperation between B cells and T cells. This was a watershed moment in the history of modern immunology. These experiments were reported in the years 1966 to 1968, and it was an exciting time.

Chapter 5

The significance of cellular cooperation in the generation of immune responses

The general agreement in the early 1970s that the activation of B cells requires or is facilitated by T cells led to two questions. What is the physiological significance of such cooperation? What might such cooperation mechanistically entail? These are related questions, and two main proposals were put forward.

The role of T cells as facilitators of antigen presentation

An interesting series of observations led to a proposal for what the physiological role of T cells might be in the activation of B cells.

We have noted that neonatal thymectomy of mice results in the adult being immunologically highly compromised. However, these mice can still produce antibody to some particular antigens. Such mice can produce as vigorous an antibody response as normal mice to certain carbohydrate antigens. Each molecule of these carbohydrate antigens contains many chemically similar sites, as they are made from repeating units. These molecules are called *polymeric antigens*. A consequence of their nature is that one molecule of a polymeric antigen can bind to many different antibody receptor molecules present on the surface of a single antibody precursor cell. It seemed that when antigen interacts with a B cell in this way, the B cell can be activated without T cells.

Moreover, artificial polymeric antigens were also shown to be able to activate B cells without the apparent presence of T cells. These polymeric antigens predominantly induce the formation of IgM antibody. Such antigens are called ***thymus-independent antigens***.

Feldman and Mitchison elaborated the view that T cells help less polymeric antigens to present an array of identical epitopes to the B cells so that these epitopes are presented in an array similar to that of epitopes naturally presented by polymeric antigens. This highly popular model, which I refer to as the Antigen Bridge Model (Figure 9), implied that T cells are not strictly required to activate B cells, but facilitate or help antibody responses. The T cells, which facilitate the activation of B cells, were referred to as ***T helper cells***. This name quickly became universal.

From a contemporary point of view, it is unclear whether there really are such thymus-independent, polymeric antigens. However, this is not such a major issue. What has become apparent is that the T helper cells do much more than present antigen to the B cell, as we shall shortly see, in a manner not envisaged by the proponents of this model, and inconsistent with a purely presenting role of T helper cells.

The single lymphocyte/multiple lymphocyte model for the antigen dependent inactivation/activation of lymphocytes

Cohn and I developed a different perspective. Our initial considerations were made in 1966.

The model we developed was motivated by two thoughts. A correct description of how antigen interacts differently with lymphocytes, to result in their activation and inactivation, should incorporate a mechanism of peripheral tolerance. This was really just an expression of the belief that the physiological significance of lymphocytes being susceptible to both activation and inactivation pathways was to achieve peripheral self-nonself discrimination.

Secondly, it seemed the proposed mechanisms of lymphocyte activation and inactivation, and thus of peripheral self-nonself discrimination, should be consistent with the Historical Postulate. Thus, the inactivation of lymphocytes specific for peripheral self-antigens should be a consequence of their presence early in development and their continuous presence thereafter, as expressed in Figure 4 of Chapter 3. We developed this model before it was generally realised that distinct types of lymphocyte exist.

Lymphocyte cooperation in context of the older literature

First came the observations of Weigle, already outlined in the previous chapter, showing that immunization with HSA can break the unresponsive state to BSA in BSA-unresponsive rabbits. There inevitably are quantitative implications of the idea that antigen-mediated lymphocyte cooperation is required to achieve lymphocyte activation. A single antigen-specific lymphocyte cannot be activated. Suppose there must be a minimum of a hundred antigen-specific lymphocytes in an animal to make antigen-mediated lymphocyte cooperation somewhat likely, and so initiate an immune response. Anti-BSA antibody precursor cells are present in BSA unresponsive rabbits, as deduced from the fact that anti-BSA antibody can be raised by immunizing with HSA. As BSA is itself not immunogenic in BSA unresponsive rabbits, the BSA unresponsive rabbits cannot harbor the hundred BSA-specific lymphocytes required to generate an antibody response. As HSA itself is immunogenic, there must be at least a hundred HSA-specific lymphocytes. Thus, HSA can activate the B cells that have a receptor able to bind to both HSA and to BSA.

PLL non-responder guinea pigs do not produce anti-PLL antibody when challenged with PLL, but they do when challenged with PLL coupled to the immunogenic carrier BSA. Again, anti-PLL antibody precursor cells must exist in these non-responder guinea pigs, as PLL antibody can be produced. The activation of these PLL-specific antibody precursor cells, on challenge with PLL-BSA, presumably occurs as the more numerous BSA-specific lymphocytes are recruited to aid the activation of PLL-specific antibody precursor cells. A similar argument can explain why B_4 unresponsive rabbits can produce anti-B antibody on immunization with A_2B_2.

An explanation of peripheral tolerance consistent with the Historical Postulate

The realization that lymphocyte cooperation is required to activate lymphocytes naturally leads to the question of what happens when antigen interacts with a single antigen-specific lymphocyte. One possibility is that the lymphocyte is inactivated. We recognized with interest that this possibility provided an explanation of peripheral tolerance consistent with the Historical Postulate, as illustrated in Figure 10. Thus, lymphocytes specific for a self-antigen would be inactivated by this antigen as they first arise, one or a few at a time, by virtue of the continuous presence of the self-antigen. In contrast, lymphocytes specific for a foreign antigen F accumulate in the absence of this antigen. When F impinges upon the immune system, it can mediate the lymphocyte cooperation required to generate an immune response. We naturally explored whether these proposals were consistent with the observations at hand.

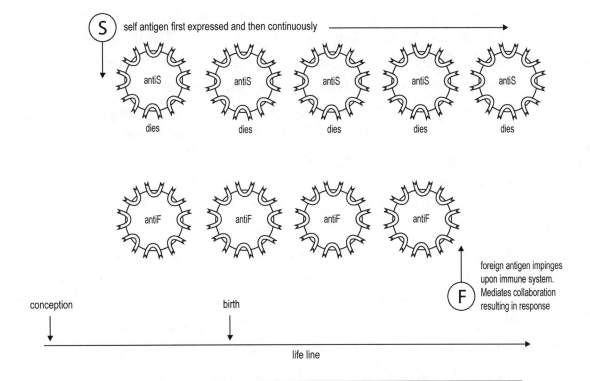

Figure 10. The One Lymphocyte/Multiple Lymphocyte Model for the antigen-dependent inactivation/activation of lymphocytes

A paradox and its potential resolution

Most of the observations listed in the last chapter under "Carrier effects" can be explained if the activation of antibody precursor cells requires antigen mediated lymphocyte cooperation. Indeed, we have seen that the first three observations can be explained on the hypothesis that lymphocyte activation requires or is facilitated by lymphocyte cooperation. The fourth and last observation cannot be explained in this manner because it is not pertinent to the requirements to generate an antibody response. This observation is directed at the requirements to generate unresponsiveness. The pertinent finding is that when B_4 is administered to B_4-unresponsive rabbits, it does not inhibit the production of anti-B antibody on a subsequent challenge with A_2B_2 a few weeks later. This, and similar findings in other systems, had been interpreted as showing that non-immunogenic molecules could not ablate immune responses. The conclusion was that non-immunogenic molecules were also **non-tolerogenic**. In the particular case just considered, it seemed that B_4 could not ablate anti-B antibody precursor cells.

This conclusion is incompatible with the idea that the mere interaction of antigen with the antibody receptors of an antibody precursor cell could inactivate the precursor cell. Was the finding wrong, or was the idea that single lymphocytes are inactivated on interacting with antigen wrong, or was there a way of reconciling the finding and the idea? I will argue that quantitative considerations are essential to resolve this issue.

Maintenance of unresponsiveness against self-antigens requires their continuous presence. We have also noted that the maintenance of an experimentally induced unresponsive state against an antigen requires the continuous presence of the antigen.

Let us consider a particular scenario. Suppose it takes four months, say 130 days, for a mouse to regain its ability to recover responsiveness against BSA, the unresponsiveness having been established by administering substantial amounts of BSA on the mouses's first day of life. Suppose the BSA antigen has been sufficiently catabolized when the mouse is thirty days old that its concentration is now too low to inactivate newly generated anti-BSA lymphocytes. Let us also suppose that one hundred anti-BSA lymphocytes are required to support an anti-BSA response. As recovery of responsiveness occurs at about 130 days of age, we can deduce, within this supposed context, that on average one new anti-BSA lymphocyte is generated per day.

Consider now the experiment carried out with B_4-unresponsive rabbits. The administration of B_4 does not noticeably affect the anti-B antibody response when the rabbits are challenged some weeks later with A_2B_2. However, A_4 is immunogenic in these rabbits, so there must be a greater number of anti-A lymphocytes than a hundred. If B_4 is tolerogenic, as we surmised, only a few anti-B antibody precursor cells need to have been regenerated on the day of challenge with A_2B_2 to generate significant anti-B antibody. As there are more than one hundred anti-A lymphocytes, A_2B_2 could activate a single anti-B antibody precursor cell. As perhaps one anti-B lymphocyte is generated per day, it is plausible there would be a substantial number of B specific antibody precursor cells at the time of the A_2B_2 challenge, even if the administration of B_4 had inactivated the anti-B antibody precursor cells. These quantitative considerations cannot but lead to some skepticism of the maxim that non-immunogenic molecules are non-tolerogenic.

Two further considerations reinforce this view. We have seen that deaggregated BSA is tolerogenic, and commercial BSA containing aggregated BSA molecules is immunogenic. Here is an example, one of several that could be drawn from the older literature, where non-immunogenic molecules are ***tolergenic***. Moreover, most of the lymphocytes that recognize commercial BSA would also most probably recognize deaggregated BSA. Thus, deaggregated BSA can ablate nearly all the lymphocytes specific for commercial BSA. This consideration provides a simple explanation of why the tolerogenicity of deaggregated BSA is so readily revealed by this experimental protocol. It would take months for recovery from the unresponsive state to take place.

Another consideration is more conceptual than observational. According to the one lymphocyte/multiple lymphocyte model, self antigens are intrinsically tolerogenic, not immunogenic. It seems that, if the maxim held that non-immunogenic molecules were also non-tolerogenic, peripheral self antigens would not inactivate their lymphocytes unless the peripheral antigens were immunogenic. It seems difficult to imagine a simple mechanism of peripheral tolerance consistent with this maxim.

These ideas represent what I now consider to be the best part of a paper Cohn and I wrote in 1968 that was published in *Nature*.[13] There were additional proposals outlined in this paper, which we later judged to be implausible. We wrote a more considered paper in 1970, which was published in *Science*.[14] I am comfortable today with all the foundational ideas outlined in this second paper.

Translating abstract ideas into concrete terms

We employed the above considerations as we attempted to understand how peripheral tolerance might be realized. Now, almost fifty years later, with so much detailed knowledge available, it seems that these abstract ideas need to be given a formulation cast in more concrete cellular and molecular terms. My immunological colleagues often request, if not demand, a molecularly detailed mechanism.

I am averse to transforming the abstract argument into a detailed cellular or molecular description and championing this mechanism, unless the transformation is compelling in being uniquely plausible. There is always the concern that a proposed particular and detailed mechanism, not unique in satisfying the abstract argument, may be found wanting. The consequences may be that people reject both the abstract considerations together with the particular mechanism proposed. Despite recognizing this hazard, it is useful to make detailed molecular proposals, as these can be more readily tested. As I do so, I hope deficiencies in particular molecular models will not lead to a loss of interest in the abstract considerations.

The general idea developed above can be summarized as the one lymphocyte/multiple lymphocyte model for the antigen-dependent inactivation/activation of mature lymphocytes. I now discuss how this principle at the level of the system might be satisfied by cellular/molecular mechanisms.

The original Two Signal Model of B cell activation

We have outlined one view as to the role of T helper cells in the activation of B cells. This role is to facilitate the presentation of an array of antigen molecules to the B cell, such that the B cell is activated by a mechanism similar to that achieved by polymeric, thymus-independent antigens. A slightly weaker form of this theory was that the T helper cells help the activation of B cells by concentrating the antigen in a manner described by the Antigen Bridge Model.

This conceptual scheme did not attempt to address how antigen might inactivate B cells. No direct communication between the T helper cell and the B cell was envisaged in the model.

We now consider how the one lymphocyte/multiple lymphocyte model for the antigen dependent inactivation/ activation of lymphocytes might bear on the mechanism by which B cell activation is dependent on T helper cells.

First, the model clearly states that the interaction of antigen with a B cell's antibody receptors can inactivate the B cell. This must mean that such an interaction is translated into a biochemical message inside the cell. Mel Cohn and I wanted a general way of referring to such an inactivating signal, and proposed that it should be called signal 1, see Figure 11. According to the one lymphocyte/multiple lymphocyte model for lymphocyte inactivation/activation, the presence of a T helper cell was critical in determining whether antigen activated or inactivated the B cell. Somehow, the B cell had to be informed that the T helper cell was close by and that its receptors were interacting with antigen. We proposed this could only be ensured if B cell activation also required the delivery of signal 2, and if the generation and delivery of signal 2 followed the recognition of antigen by the T helper cell. We proposed that signal 2 is mediated by short range molecules produced by T helper cells, which bind to receptors on the B cell, and/or by a membrane/membrane interaction between the B cell and T helper cell. Subsequent findings showed that the signal 2 delivered by T helper cells to B cells is mediated by both mechanisms.

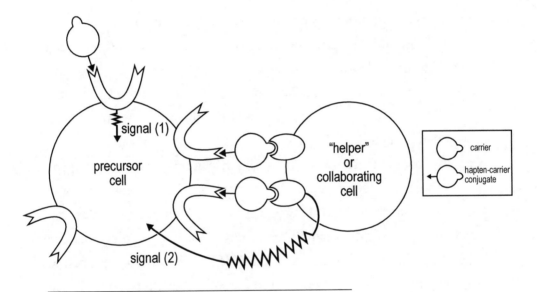

Figure 11. The original Two Signal Model for Lymphocyte Activation

The molecular nature of signal 1 initiating B cell inactivation

A number of observations support the proposition that the interaction of antigen with a B cell can result in the cell's inactivation in the absence of antigen-specific helper T cells. Such reports are generally taken as evidence for our Two Signal Model of B cell activation, and that the sustained generation of signal 1 alone results in B cell inactivation.

In addition, other observations over the years have been generally interpreted as showing that signal 1 will be generated only when antigen interacts with *membrane immunoglobulin (mIg)* receptors of the B cell, resulting in receptor aggregation. It is almost universally accepted that such *cross-linking of receptors* is required to generate signal 1. This *cross-linking model* means that an antigenic entity, able to generate signal 1, must have identical (or similar) epitopes on its surface, and with a mutual disposition in space, such that these epitopes are able to crosslink the mIg receptors in the required manner. I have for decades been most uneasy with this conclusion for two major reasons.

First, it is known experimentally that monomeric antigens, not expected to have identical epitopes, are potent in their ability to generate unresponsive states. A classical example already discussed is deaggregated BSA. Moreover, the inference from such observations is that a monovalent interaction of antigen with mIg receptors of the B cells can generate signal 1. This makes biological sense in that many self-antigens are monomeric. Second, these observations and considerations are in accord with another plausible proposition. It is unlikely that there can be any requirement on a self-antigen for it to inactivate precursor cells beyond its ability to bind to antigen specific receptors on precursor cells. If there were such further requirements that could not be satisfied by a substantial fraction of self-antigens, the inactivation of anti-self precursor cells could only be partially accomplished. This would surely make autoimmunity much more prevalent. I identified such crosslinking of mIg by self-antigens as likely being too restrictive, and so implausible, as a requirement to inactivate B cells.

In view of these considerations, I have tried over the years to imagine mechanisms by which monovalent binding of antigen to mIg molecules could result in the generation of an intracellular signal, and have come to believe such mechanisms are realizable and therefore plausible. In the last fifteen years or so Michael Reth and colleagues have provided provocative evidence against the cross-linking model.[15] They suggest that unbound mIg are in a mutually interacting, non-signaling state, and that antigen, on interacting with mIg in a monovalent fashion, is able to disperse these molecules leading to intracellular signals in B cells.[16] I only recently became aware of this research and was delighted and relieved. This evidence, if it stands up to scrutiny, removes what for me was the biggest conceptual problem with the Two Signal Model in the context of the inactivation and activation of B cells.

The activation of T cells and the priming problem

Immunologists were well aware in the late 1970s of the difference between an antibody precursor cell, which produces membrane-bound antibody receptors, and the antibody-secreting cells it gives rise to when activated. The ***precursor cell***, when activated, gives rise to antibody-secreting ***effector cells***.

It was clear in 1970 from Mitchison's studies that T helper cells could be activated to express memory. The question arose as to whether the activation of a B cell could be facilitated by the constitutive activity of T helper cells, or whether effector T helper cells had to be generated from precursor helper T cells before antigen could optimally activate B cells. Cohn and I opted for the latter view.[14] Besides the evidence that T helper activity could be induced and that T helper cells could express immunological memory, as demonstrated by Mitchison's observations, it seemed likely that the activity of T helper cells was of pivotal importance. The possibility that precursor helper T cells have to be activated to be optimally effective as helpers allows their helper activity to be more readily regulated than if T helper cell activity is constitutive. Indeed, we proposed that the activation of precursor T helper cells, to generate effector T helper cells, also required T helper cell collaboration. We shall later return to consider this and alternative ideas.

One last point should be made for clarity. We also proposed that effector T helper cells would be required to optimally facilitate the antigen-dependent activation of precursor T helper cells. In other words, effector T helper cells are required for antigen to optimally facilitate their further generation. This proposal leads to the ***priming problem***: what is the source of the first effector T helper cells?

Priming problems occur in other areas of biology. Life is needed to beget life, so how did life originate? Ribosomes are the sites of protein synthesis, and yet ribosomes are substantially composed of ribosomal proteins. Where did the first ribosomal proteins come from? We shall not address the priming problem at this particular juncture. However, I thought it helpful to note its existence so the reader can relax, knowing we will later return to it.

Chapter 6

The roles of the molecules of the major histocompatibility complex

There is no doubt that a recognition of the central role of the molecules, defined by the genetic region known as the *major histocompatibility complex (MHC),* has led to one of the most exciting and unifying sets of interrelated immunological phenomena and ideas of the last fifty years. Moreover, the nature of the progress has been so compelling that many previous puzzles have been solved, and I think the currently envisaged roles of MHC molecules are bound to remain as the field advances. I therefore try to describe these advances as succinctly as possible. Again, the few pages devoted to this subject is not reflective of their importance, but rather a measure of substantial and consolidated progress. Readers interested in pursuing this topic at greater depth can consult Janeway's article, see reference 17.

Peculiar Properties of the MHC genes

The MHC was first defined by studies on rejection of skin grafts. It was found that grafts between two genetically disparate mice were rejected in either less than twelve days or in more than twenty. It was further found that when rejection took twelve days or less, the donor of the graft was genetically different from the recipient in a genetic region known as the MHC complex. Thus, differences at the MHC, when they exist, are dominant in determining the speed of the rejection process.

Genetic loci are sites at a defined position on a chromosome where different genes are found among members of the population, or within an individual, if this individual has different genes on their maternal and paternal chromosomes. Alternative genes at a particular genetic locus are referred to as **alleles**. Usually there are only a few different alleles at a given genetic locus among the population, and they code for poly-peptide chains that differ by an amino acid or two. The several MHC loci, situated close together on a chromosome, differ from this standard picture in two respects. There can be a hundred different alleles within the population at one genetic locus. Moreover, when two alleles are examined belonging to the same genetic locus, it is often found that the polypeptide chains they code for can differ in their amino acid sequence at thirty different positions. Thus, the diversity of proteins encoded by the MHC has long been recognized as exceptional. It is natural to hope that an understanding of the function of MHC mol-ecules might provide clues as to the evolutionary pressure that results in such diversity.

The several genetic loci that exist within the MHC, being close together on the chromosome, tend to be inherited together. All the alleles on a maternal, and all the alleles on a paternal chromosome, are inherited by offspring as a block about 99% of the time.

Two kinds of MHC molecules

There are two types of MHC molecules, called class I and class II MHC molecules.

Chemically, class I MHC molecules consist of two polypeptide chains. One chain, β2 microglobulin, is invariant. The second chain is highly variable. In the mouse, there are K and D class I MHC mol-ecules. The variable polypeptide chains of K class I MHC molecules are coded for at the K locus, and the variable polypeptide chains of D class

I MHC molecules at the D locus. Both types of class I MHC molecules are found on the surface of all nucleated cells, in other words, on all cells except enucleated red blood cells. In humans, there are three types of class I MHC molecules, whose variable chains are encoded at three different loci. Wild type mice usually inherit different K and D molecules from their two parents, and so their cells express four different class I MHC molecules.

Class II MHC molecules each consist of two highly variable chains. In the mouse, there are I-A and I-E class II MHC molecules, made up respectively from Aα and Aβ, and Eα and Eβ chains. Four corresponding genetic loci code for these four chains. Many different alleles are found at these four loci. In contrast to class I MHC molecules, class II MHC molecules are found primarily on specialized cells. These cells are specialized in having means for taking up external antigens from outside the cell into intracellular cellular compartments. Macrophages thus express class II MHC molecules. It turns out that B cells can take antigen that binds to their receptors into the cell by a process called *endocytosis*, and the antigen molecules land up in what is called the endocytic compartment of the B cell. B cells also express class II MHC molecules.

Overview of the function of class I MHC molecules

It appears that class I MHC molecules have a central role as part of a spy system that allows the immune system to monitor what is going on inside a cell.

To set the scene, consider an intracellular parasite that tries to partially take over host machinery for its replication, but does not allow the insertion of any of its own molecules into the external membrane of the infected cell. The immune system would appear at first glance to have no means of distinguishing the infected from an uninfected cell, and would therefore be unable to protect against this invader. Consider another scenario. Suppose some mutations occur in a cell that affect the function of internal proteins that control the cell cycle; the mutated cell grows in an unrestrained fashion and is cancerous. Again, it would appear that the immune system has no way of distinguishing this cell from a benign cell. These examples are perhaps fanciful. However, they provide instances that allow one to imagine the potential importance for the functioning of the organism of the immune system's ability to detect what is going on inside a cell. It appears that class I MHC molecules play a pivotal role in this surveillance. In understanding this role, it is important to appreciate several features of protein structure, and one structural feature of MHC molecules

Intact proteins contain polypeptide chains that are long, often of more than a hundred amino acids in length. Consider a section of a long chain that consists of a stretch of ten amino acids. This section belongs to a large structure and so is held in place by interactions with several neighboring parts. The large protein has a definite shape. When a polypeptide chain is cut up into oligo peptides, say about ten amino acids in length, the peptides are much more flexible. They can exist in an extended conformation in which the two ends are as far apart from each other as they can be.

As already noted, polypeptides are made up of twenty amino acids. All amino acids are identical in structure except for the side chain that is characteristic of each. When the amino acids condense to form a polypeptide, the parts that are shared form the links between the individual amino acids, so all polypeptides have a chemically identical backbone.

All MHC molecules have a groove, as shown in Figure 12. This groove has intrinsic affinity for the common backbone of oligo peptides and is of a size to fit them in the extended form of nine or ten amino acids in length. However, there are variations between different MHC molecules at the sites where the side chains of the amino acids of the oligo peptide must be accommodated if there is to be effective binding, so only certain oligo peptides will bind well to a particular groove. Interestingly, the variability between different class I MHC molecules is concentrated around the floor, the walls, and the lips of the groove. Thus, different class I MHC molecules bind different peptides.

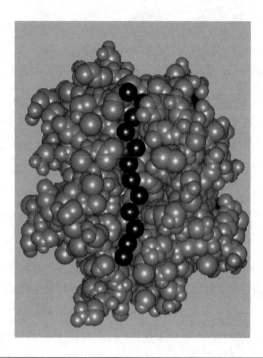

Figure 12. Looking down onto the peptide-binding groove of a peptide/MHC complex. The black spheres represent the bound peptide, the grey spheres the MHC molecule.

The function of class I MHC molecules is related to these structural characteristics. A sample of the proteins synthesized in a cell is cut up into oligo peptides of variable size, but about ten amino acids in length. These oligo peptides are in a compartment of the cell where they have the opportunity to bind to newly synthesized class I MHC molecules. Some of the oligo peptides will bind to the grooves of the class I MHC molecules, and these peptide/MHC complexes are then transported to and inserted into the external membrane of the cell. In this way, the spectrum of proteins inside the cell is represented by peptide/MHC class I complexes expressed on the cell's surface. A normal cell will express a range of peptides derived from self-proteins. A virally infected cell will express many of the same self-peptides but, in addition, it will express some peptides derived from viral proteins.

The activation of CD8 T cells

Infection of a mouse with a virus usually results in the generation of T cells that are cytotoxic for, or can lyse, host cells infected with the virus, but are unable to lyse uninfected host cells. These cytotoxic T cells have considerable specificity, and usually cannot lyse host cells infected with a virus unrelated to the virus used for infection. Such *cytotoxic T lymphocytes (CTL)* are important in protecting against viral infections and against cancers. The receptor of these virus-specific CTL recognize peptides derived from viral proteins bound to class I MHC molecules expressed on the surface of virus-infected cells.

In the mid-1970s, Peter Doherty and Rolf Zinkernagel made a remarkable discovery for its time. These CTL did not lyse cells obtained from other strains of mice and infected with the virus unless these strains shared class I MHC molecules with the host in which the CTL were generated. These findings were made well before the chemical and three-dimensional structure of MHC molecules had been elucidated, or genes encoding the T cell receptor (TcR) were characterized. We also now know the three dimensional structure of the T cell receptor and how it interacts with the peptide/MHC complex in space. This interaction involves recognition by the TcR of some of the amino acid side chains of the oligo peptide bound to the groove of the class I MHC molecule, as well as the lips of the groove. As the TcR recognizes in this case the peptide/class I MHC complex, the T cells are said to be class I MHC-restricted.

These cyototoxic T cells also express the CD8 surface molecule. The CD8 molecule binds to an invariant part of class I MHC molecules, and this interaction increases the ability of CTL to bind to target cells. These insights give a vivid picture of the role of CTL in providing pro-tection, and of class I MHC molecules as central to the mechanism by which the immune system gains information about the intracellular proteins of cells.

To provide context, it is useful to note that the sustained activation of naïve CD8 T cells, which can be activated to yield cytotoxic CD8 T cells, requires CD4 T helper cells. In their absence, antigen can inactivate the CD8 T cells. These general observations fit with the one lymphocyte/multiple lymphocyte model for the inactivation/activation of lymphocytes. Thus, these and previously described observations have led to the recognition that T helper cells play a pivotal role in determining whether antigen activates B cells and CD8 T cells. This is why there is such a research focus on what determines whether antigen inactivates or activates T helper cells.

The activation of CD4 T helper cells: antigen processing and presentation

The expression of class II molecules occurs primarily on cells that have some means of taking in extracellular proteins and other molecules. In the case of macrophages, external antigens can be taken up by phagocytosis, and initially the antigen lands up in a membrane-enclosed compartment called the phagosome. These phagosomes fuse with another membrane-enclosed compartment called a lysosome to produce a phagolysosome. The lysosome and phagolysosome contain *proteolytic enzymes* that degrade protein antigens into oligo peptides. Newly synthesized class II MHC molecules are directed to this compartment, and some of these oligo peptides bind to the grooves of class II MHC molecules. The peptide/class II MHC complexes are then transported to and inserted into the external membrane of the cell. Thus, peptides derived from internally synthesized proteins, so called *endogenous antigens*, are presented by class I MHC molecules, and peptides derived from proteins synthesized outside and taken up by the cell, so called *exogenous antigens*, are presented by class II MHC molecules. The antigen-specific receptor of a T helper cell recognizes a peptide bound to class II MHC molecules. Again, the TcR recognizes some of the side chains of the oligo peptide bound to the groove, as well as to the lips of the groove. All T helper cells express the CD4 surface molecule. The CD4 molecule binds to an invariant part of class II MHC molecules, and this interaction again facilitates the interaction of the TcR with its ligand.

The taking in of exogenous antigens by macrophages, and the subsequent degradation of the antigen, is referred to as **antigen processing**, and the final event, when peptide/MHC complexes are expressed on the external membrane, is called **antigen presentation**. The intact antigen, from which peptides are derived by processing, is called the **nominal antigen**. Suppose we immunize a mouse with the antigen **ovalbumin (OVA)**, which is highly present in egg white and so often employed because of its ready availability. This antigen may be processed and be presented by macrophages. Such processing and presentation is an essential step in the activation of CD4 T cells, as it allows the TcR of the CD4 T cell to bind its ligand. We shall later discuss at some length the significance of, and the envisaged requirements, to activate CD4 T cells. We just note now that immunologists have found a convenient way of referring to these CD4 T cells that recognize OVA peptide/class II MHC complexes. They refer to them as CD4 T cells specific for the **nominal antigen** ovalbumin. The word nominal before the word antigen indicates that the TcR does not recognize the antigen itself but rather a peptide derived from the nominal antigen, which is itself bound to the groove of an MHC molecule.

The MHC-restricted model of the B cell/ CD4 T helper cell interaction

We have already seen that whether antigen activates or inactivates a B cell depends upon the presence or absence of T helper cells. It was natural for over a decade to envisage the interaction between a B cell and a T helper cell in terms of the Antigen Bridge Model (Figure 9). Moreover, this model accounted for a most important feature of the interaction between the B cell and the T helper cell first delineated by Mitchison and Rajewsky. Mitchison employed his experimental model to analyze the requirements to generate secondary anti-hapten antibody responses. A primed helper T cell specific for an antigen Q will only help the activation of an h-specific memory B cell in the presence of h-Q, and not in the presence of Q and h-R, where R is a protein carrier that does not crossreact with Q. In other words, the fruitful interaction of Q-specific primed T helper cells and h-specific memory B cells does not merely require the presence of h and Q, but that they are present and physically linked to one another.

This is a critical point, particularly in the context of the Two Signal Model of B cell activation. Consider the consequences, within this context, if linked recognition was not required. The activation of a B cell, specific for a peripheral self antigen pS, could be facilitated by a foreign, non-crossreacting antigen F, in the presence of T helper cells specific for nominal F. The inactivation of the pS-specific B cell could be interfered with by an immune response to an irrelevant antigen. This would seem to be a physiologically unlikely situation. It seems that evolution would have avoided such interference. This lack of interference can be achieved if the B cell/T helper cell interaction is mediated by a mechanism involving the operational recognition of linked epitopes, as experimentally found. We refer to mechanisms, whether actual or proposed, that avoid interference as satisfying the ***Principle of Non-Interference***.

The fact that the receptors of T helper cells do not bind directly to the nominal antigen means the Antigen Bridge Model cannot be correct. Moreover, consider the activation of an anti-hapten B cell by the antigen hQ, facilitated by T helper cells specific for the nominal antigen Q. How can the operational recognition of linked epitopes be realized when Q must be processed into peptides before the receptors of the T helper cell can recognize their ligand, thus severing the linkage between the hapten and carrier known to be required to achieve the activation of the B cell? Antonio Lanzavecchia developed the ***MHC-restricted B cell/T helper cell model*** of B cell activation, which solved this conundrum, see Figure 13.

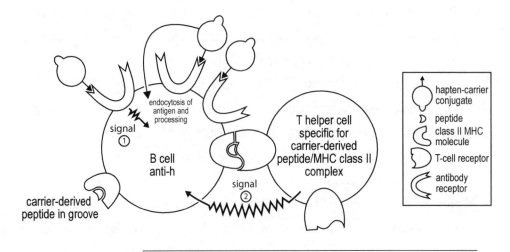

Figure 13. The MHC-restricted model of the B cell/ T helper cell interaction

Lanzavechhia demonstrated that the anti-hapten B cell can bind hQ via its Ig receptors, resulting in the uptake of hQ into the B cell, the processing of Q in the membrane enclosed endocytic compartment. This process then leads to the presentation of peptides, derived by processing Q, by class II MHC molecules in the outer membrane of the B cell. This presentation of Q peptides now allows a Th cell, specific for the nominal antigen Q, to bind to the Q-derived peptide/class II MHC complexes and deliver signal 2. Note that the carrier Q will only be taken up by the hapten-specific B cell if the hapten is linked to Q. Thus, for Q-specific T helper cells to help the activation of h-specific B cells, h must be attached to Q. This model thus explains how the interaction between the B cell and T helper cell involves the ***operational recognition of linked epitopes***.

Chapter 7

The central role of antigen presenting cells

The discovery of costimulation and the activation of T cells

The unusual strength of MHC antigens

We have already noted that grafts bearing foreign MHC antigens are rapidly rejected. This rapid rejection is due to strong immune responses against MHC antigens. A major impetus to understand the nature of these strong immune responses was to be able to subvert them, and so achieve long-term transplantation across MHC barriers.

According to the Clonal Selection Theory, antigen can interact in some manner with antigen-specific lymphocytes to cause their activation and multiplication. However, antigen-specific lymphocytes are scarce and so, when a foreign antigen is added to a cellular population containing a substantial fraction of unprimed lymphocytes, it is difficult to detect whether scarce lymphocytes multiply in response to antigen stimulation. There is one exception to this statement. When spleen cells from one strain are mixed with spleen cells from an MHC incompatible strain, there is massive lymphocyte proliferation. Immunologists often treat one population with a drug that blocks cellular multiplication, so it is clear that the lymphocyte proliferation detected reflects the stimulation and resulting multiplication of the lymphocytes of the untreated population. This experimental setup is referred to as a *one-way mixed lymphocyte response*. The drug-treated lymphocytes are referred to as the **stimulators**, and the dividing lymphocytes are referred to as the **responders**. The stimulating spleen cell population can be replaced by different cell populations or cell lines, bearing the same stimulating MHC antigens.

The Lafferty/Cunningham Model of T cell Activation

Lafferty and Cunningham noted in the mid-1970s that not all cells with strong expression of MHC antigens are strong stimulators. They respectively referred to strong and poor stimulators as having an S^+ and an S^- phenotype. They proposed that cells of both phenotypes were able to interact with the receptors of anti-MHC T cells to generate a signal, referred to as signal 1. They proposed that S^+ cells could in addition generate a second signal, by virtue of expressing a molecule on their surface complementary to a receptor on the T cell. In time, the molecule on the S^+ cell was designated as the costimulatory molecule, with its counter receptor expressed by the T cell.

These findings were seminal. They convinced immunologists of the value of a two-signal framework when considering the activation of T cells. Lafferty and Cunningham gave Bretscher and Cohn credit for first suggesting that lymphocyte activation required two signals. I consider this credit somewhat unfortunate. Although signal 1 is the same in the Bretscher/Cohn model of lymphocyte activation and in the Lafferty/Cunningham model for the activation of T lymphocytes, the second signal is mechanistically different. This difference has profound physiological implications. The whole issue became, to my mind, a most unfortunate confusion in the field. The differences between these two-signal formulations raise foundational issues to which I shall return.

The CD28/B7 Model for Costimulation

Over a decade after the Lafferty/Cunningham formulation, observations by Quill and Schwartz led to a more detailed model of CD4 T cell activation. The activation of a CD4 T cell was envisaged to require the generation of two signals, signal 1 generated upon the interaction of the TcR with its ligand, peptide/class II MHC complexes on the surface of an antigen presenting cell. A second *costimulatory signal* is generated when a surface molecule of the CD4 T cell, the CD28 molecule, interacts with B7 molecules, present on the antigen-presenting cell. The B7 molecules are referred to as *costimulatory molecules*, and CD28 as its counter receptor. This model, see Figure 14, is the first of several in which the absence/presence of costimulatory signals determines whether antigen inactivates or activates CD4 T cells. According to this model, in contrast to others to be discussed later, the costimulatory molecules are constitutively expressed by the antigen presenting cell, and so this model can be conveniently referred to as the *Constitutive Model*.

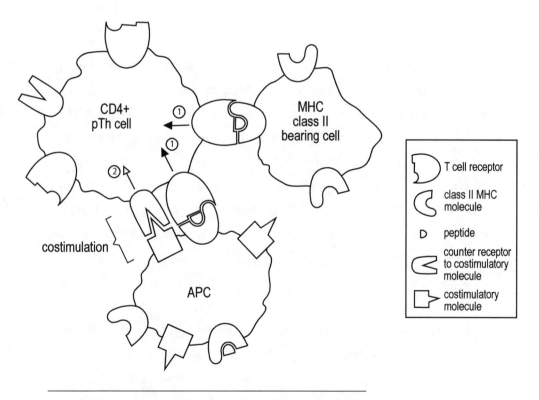

Figure 14. The constitutive costimulatory model of CD4 T cell activation

Different types of antigen presenting cells

It is perhaps not surprising, particularly in hindsight, that APC play a central role in activating T cells, both CD4 and CD8 T cells. It might be thought that this role reflects their phylogenetically ancient function as phagocytes. Phagocytes have diverse receptors to aid in the uptake of foreign matter, a function suitable for a cell involved in the earliest events of initiating immune responses. The role of antigen presenting cells now appears central through their three different functions: processing of antigen, presentation of antigen, and now costimulation.

There are at least three different types of *antigen presenting cells*, or **APC.** I argue below that the subtle differences between different types of APC is of foundational importance. I shall outline the characteristics of the three types, stressing their differences, as they likely reflect profoundly different physiological roles.

Macrophages

The first type to be recognized was the macrophage. We have already seen that it is difficult to detect the antigen-dependent proliferation of unprimed lymphocytes to antigens other than to MHC antigens. However, when a population of lymphocytes primed against ovalbumin (OVA), such as spleen cells from an OVA-immunized mouse, is incubated with OVA, lymphocyte proliferation can be readily detected. Immunologists were happy with this finding. It seemed to confirm the idea that memory is partly due to a higher frequency of responding lymphocytes among memory lymphocytes than are present among unprimed lymphocytes. Moreover, such proliferation provided an assay for recognition of antigen by lymphocytes.

In trying to ascertain the nature of the responding cell, researchers found that another cell type besides lymphocytes is required to support proliferation. It was a macrophage by several criteria, including its ability to stick to plastic and engulf foreign matter. Moreover, purified, OVA-primed CD4 T cells could be stimulated by OVA to proliferate in the presence of purified macrophages. Thus, this assay could be employed to examine the requirements for antigen to be recognized by CD4 T cells. Indeed, this system was exploited to delineate the steps involved in antigen processing and presentation.

Two interesting observations made in this system provide a context for further developments. First, OVA could be replaced by a protease digest of the protein, resulting in a mixture of oligo peptides about ten to twenty amino acids in length, in causing proliferation of OVA-primed CD4 T cells. This finding demonstrates that CD4 T cells recognize antigen in a different manner from the way antibody does. Secondly, it was found that such proliferation could be blocked by antibodies specific for the class II MHC molecules present on the surface of the macrophage. This is anticipated if the receptors of the CD4 T cells recognize peptide/class II MHC complexes.

The ability of OVA peptides, when added to cultures of OVA-primed CD4 T cells to stimulate these cells to divide, is envisaged to occur in the following manner. Some of the resident peptides, bound to the grooves of class II MHC molecules on the surface of the macrophages, come free of the groove. Some of the OVA peptides can then bind to the empty grooves of the class II MHC molecules to take their place. This is consistent with the half-life of peptide/class II MHC complexes being of the order of twenty-four hours. Thus, the OVA peptides can be presented by macrophages following the process of peptide decoration from the outside. This appears to be an artificial, non-physiological process.

A critical point is that macrophages can be exposed to intact (non-digested) ovalbumin, thoroughly washed after a couple of hours, and then these antigen-pulsed macrophages can stimulate the OVA-primed CD4 T cells to proliferate without any further addition of antigen. These macrophages can thus process and present antigen. They also express B7 costimulatory molecules on their surface.

Dendritic cells

A second type of APC is called a dendritic cell, as it possesses many branch-like structures emanating from its cell body, *dendra* being the Greek for branch. These structures increase the surface area of the cell, increasing its ability to capture and phagocytose antigens. These cells, either as immature or mature dendritic cells, exist in various parts of the body, but are particularly prominent just under the skin and can form an impressive net of immature dendritic cells, see Figure 15.

Immature dendritic cells give rise to mature dendritic cells, perhaps spontaneously, but the maturation process is certainly accelerated under certain conditions. The immature dendritic cell is highly phago-cytic, bears some but few class II MHC molecules on its surface, as well as few B7 costimulatory molecules. When it matures, four of its charac-teristics change. First, it loses its phagocytic capacity. It also expresses more class II MHC molecules on its surface. As it no longer takes in exogenous antigen, the peptides its class II MHC molecules present are derived from the antigens the dendritic cell took up in its imma-ture state. Third, it expresses greater levels of costimulatory molecules. Fourth, it migrates to the nearest draining lymph node. Here it can present antigen to naïve CD4 T cells and initiate an immune response.

The process of dendritic cell maturation can certainly be sped up under certain circumstances. Thus, immature dendritic cells have various ***pattern recognition receptors (PRP)*** that can be triggered by ***pathogen-associated molecular patterns (PAMPs)*** to accelerate or result in maturation. One example of an effective PAMP is the lipo-polysaccharide of gram-negative bacteria. I should state two alternative views. One is that dendritic cell maturation is completely dependent upon the presence of an appropriate PAMP. The other view is that, though PAMPs can certainly accelerate the maturation process, it can occur in their absence, albeit at a slower rate. The importance of these alternatives will become apparent shortly.

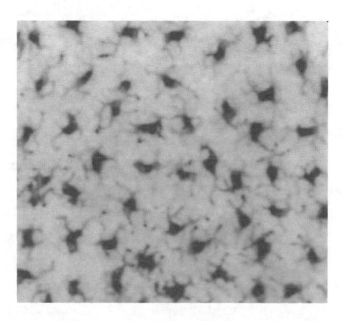

Figure 15. The dark-stained cells are dendritic cells just under the surface of the skin forming a network of sentinel cells. Note the dendra, the prominent branches of these cells.

This picture of dendritic cell maturation makes eminent biological sense. Consider a skin abrasion through which bacteria gain access to the body. The immature dendritic cells at this site of damage, see Figure 15, will phagocytose the bacteria. The bacteria, bearing diverse PAMPS, will facilitate the maturation of the dendritic cells. The mature dendritic cells migrate to the draining lymph node, taking a representation of the antigens present at the site of their maturation, including bacterial antigens. Note a difference between dendritic cells and macrophages. Mature macrophages can both phagocytose and present new antigens, and can thus act as APC for those antigens recently phagocytosed. Mature dendritic cells are not phagocytic, and do not rapidly replace the class II MHC molecules expressed on their cell surface. These features suit their role of stably reporting on the variety of antigens present at their site of maturation to T cells in secondary lymphoid organs.

Macrophages are present systemically. If you have a systemic infection that has invaded many parts and interior organs of the body, macrophages are needed to present the antigens of the pathogen to CD4 T cells in secondary lymphoid organs. Depletion of macrophages undermines immune responses to systemic infections.

The role of B cells as APC

The third type of APC is the B cell. The naïve B cell is only efficient in taking up antigens that can bind to its antibody receptor. A B cell can present peptides derived from this antigen and, as we have seen, such presentation is central to the activation of the B cell by CD4 T helper cells. Interestingly, the resting B cell expresses low levels of B7 costimulatory molecules but their expression is upregulated when the B cell is activated by CD4 T helper cells. This upregulation is consistent with the idea and with the observation that B cells can act as antigen-specific APC.

Chapter 8

Peripheral tolerance of CD4 T cells

As already discussed, observations show that CD4 T cells facilitate the antigen-dependent activation of B cells and that prolonged exposure to antigen, in the absence of T helper cells, leads to the inactivation of the B cells. The evidence is strong that helper T cells are required in several circumstances to activate naïve CD8 T cells and to generate memory CD8 T cells. In addition, when an antigen interacts with these CD8 T cells in the absence of CD4 T helper cells, the antigen can inactivate these CD8 T cells. There also are reports that antigen can activate CD8 T cells to some extent in the absence of CD4 T cells. It is unclear whether this activation of CD8 T cells involves the antigen-mediated interaction between CD8 T cells themselves, in a manner consistent with the one lymphocyte/multiple lymphocyte model for the inactivation/activation of lymphocytes. However, it is clear that CD8 T cells have some of the features of CD4 T cells essential to their helper function, and that their mutual interaction can sometimes facilitate their activation.

All these observations support the generalization that in most cases the antigen-dependent activation of lymphocytes, other than naïve CD4 T cells, requires CD4 T cells. In the absence of CD4 T cells, antigen can inactivate these lymphocytes. These circumstances have naturally led to a focus in the immunological community over the last three decades on the question of how antigen interacts differently with naïve CD4 T cells to result in their inactivation or in their activation. We discuss three foundational models.

The context for the three models of CD4 T cell activation

It is helpful to recall the broad evolution of ideas concerning CD4 T cell activation that were prominent in 1989, when Janeway first proposed his highly influential model.[18]

It was known that APC are required to present antigen to naïve CD4 T cells in order to activate them. Activation of the CD4 T cell was believed to require the generation of a T cell receptor-mediated signal when the TcR interacts with peptide/class II MHC complexes, which is referred to as signal 1. The activation of the CD4 T cell was believed to require, in addition, the generation of a costimulatory signal, the prototypical signal involving an interaction between the APC's B7 costimulatory molecules and its receptor, CD28, constitutively expressed on the CD4 T cell. Studies had led to the idea that the generation of signal 1 alone results in inactivation of the CD4 T cell, as we have seen. For example, it is possible to block the generation of the costimulatory signal by adding a ligand that binds to the APC's B7 molecules with high avidity, so preventing the interaction between CD28 and these B7 molecules and so preventing the generation of the costimulatory signal. Under these circumstances, the CD4 T cells are not activated. They become inactivated over time, in the sense of not being susceptible to further activation signals upon antigen exposure. It is worthwhile to make two points at this juncture.

All three models we discuss share the idea that there is a crucial costimulatory signal that is required for CD4 T cell activation. In the absence of its generation, the antigen inactivates the CD4 T cell. How this crucial costimulatory signal is generated, and how its delivery is controlled, are therefore fundamental questions. The three models differ in what they propose.

The second point is that there is a formal and acknowledged similarity between Janway's 1989 model and the Two Signal Model that Cohn and I proposed in 1970. In both, the generation of signal 1 alone, in the absence of the generation of the critical costimulatory signal (Janeway's 1989 model), or of signal 2 (our 1970 model), the CD4 T cell is inactivated. The CD4 T cell is only activated when the costimulatory signal or signal 2 is generated. However, though formally somewhat similar, they can have profoundly different physiological consequences. The critical feature of the signal 2 in our 1970 formulation is that its delivery to a target lymphocyte is dependent upon the recognition of antigen by an antigen-specific helper lymphocyte. This is a point to which we shall return.

Janeway's Infectious/Non-Infectious Model for the Activation/Inactivation of CD4 T cells

Janeway was struck by the fact that, in various classical studies with such antigens as BSA, in which purified vertebrate antigens are employed to examine the requirements to induce immune responses, immunologists need *adjuvants*. Adjuvants are substances with which antigens are mixed and then injected into an animal or into people to generate measurable or stronger immune responses than if the antigen was administered alone. In animals, the most commonly used and one of the most powerful adjuvants is **Complete Freund's Adjuvant** or **CFA**. Its chemical composition is complex, but essentially consists of mineral oil mixed with dead mycobacteria. Janeway famously referred to the need to use this adjuvant to obtain vigorous immune responses to purified proteins as "the immunologists' dirty little secret". He proposed that our foundational concepts were distorted by our neglect of this need, and consequently made a proposal. Remarkably, it seemed to be in accord with subsequent findings of what is now known about the role of dendritic cells as antigen presenting cells. These findings naturally increased the attractiveness of his proposal.

In a nutshell, Janeway suggested that APC must be activated to express costimulatory molecules to a level where they can activate CD4 T cells, and that such activation requires a PAMP to interact with one of the APC's PRR. In the absence of the PAMP, the APC will present the antigen with insufficient costimulatory signals to achieve activation, so inactivation of the CD4 T cell occurs, see Figure 16.[18] Janeway realized the radical implications of his proposal. It means that, at the level of CD4 T cells, the immune system does not distinguish self from nonself, but rather infectious entities containing appropriate PAMPs from non-infectious entities. As CD4 T cells are essential for initiating immune responses, it can be fairly stated that his view implies that the immune system itself discriminates between infectious and non-infectious entities, rather than nonself from self.

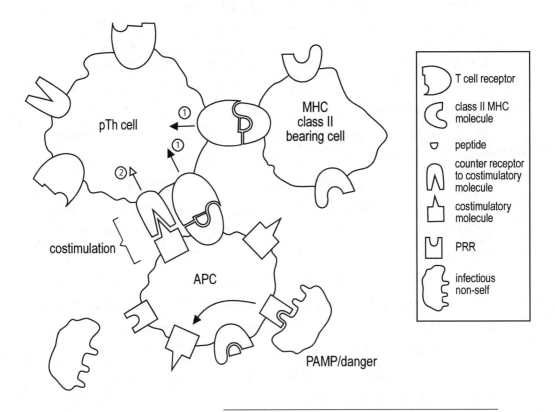

Figure 16. The PAMP/Danger Model of CD4 T cell Activation

A further series of findings gave substance to Janeway's view. Steinman and colleagues targeted antigen to dendritic cells by ingenious means. Most dendritic cells are in an immature state, and it was found that such targeting, resulting primarily in antigen presentation by immature dendritic cells, led to the inactivation of the corresponding CD4 T cells. Thus, according to this view, in the **steady state** and in the absence of an infection, inactivation of CD4 T cells occurs.

Matzinger's Danger Model

About five years after Janeway first made his proposal, Matzinger formulated the Danger Model[19], which addressed in part a weakness of Janeway's formulation. It is clear that vertebrate antigens, not expected to contain PAMPs, can be immunogenic. One much explored example would be transplants of foreign grafts within a species, an experimental model for human transplantation. Matzinger proposed CD4 T cells are activated when the body senses danger. In her current formulation, I believe Matzinger postulates those situations envisaged by Janeway as well as others, for example when there is injury, as inevitably involved in transplantation of allografts. She coined the term "danger" to provide a collective way of referring to such situations, and her proposal is referred to as the Danger Model. Janeway's and Matzinger's models are distinct, but have some features in common. When I wish to discuss these common features, I refer to them as the PAMP/Danger Model, see Figure 16. These two models are the prevalent contemporary models and have inspired enormous interest and fascinating investigations.

The Two Step, Two Signal Model

About ten years after Janeway proposed his model, I published a model that attempted what was then a contemporary formulation of the one lymphocyte/multiple lymphocyte model for the inactivation/

activation of CD4 T cells. This formulation took account of the new information and developments that had accumulated in the thirty years since the original proposal.

Cohn and I had postulated in our 1970 article that the activation of precursor T helper cells also required CD4 T helper cells. Studies by my students and myself over the years support this idea and further characterized the nature of the interaction. I outline these studies as they allow a more concrete and focused exposition.

Some experimental findings pertinent to CD4 T cell activation

Our findings were three fold. First, the activation of target CD4 T cells, specific for a nominal antigen Q, could be facilitated by helper CD4 T cells specific for a non-crossreacting nominal antigen R, in the presence of the conjugate Q-R. This facilitation of the activation of Q-specific target CD4 T cells by R-specific helper CD4 T cells required not merely the presence of Q and R but that they are physically linked. Second, we found that the ability of in vivo primed R-specific CD4 T cells to act as helpers was sensitive to radiation shortly after priming with R, but this helper activity became resistant about four days after priming. This finding led us to propose that radiation resistant effector T helper cells are generated when precursor T helper cells are activated to divide and their progeny differentiate into efficient effector helper CD4 T cells. Lastly, our observations suggested that the activation of R-specific precursor T helper cells is itself also facilitated by cooperation between CD4 T cells.

These studies certainly influenced me, and were carried out to test ideas that the studies supported, once completed. However, I find isolated observations to be only suggestive, and a framework can become compelling only when it provides a coherent picture of diverse observations in a way that makes biological sense. The tentativeness I describe here follows an awareness that all deductions are made within a set of assumptions, some of which may be later found to be wanting.

The model

The observations I have just outlined were made and reported some years before Janeway's Infectious/Non-Infectious Model for the activation and inactivation of CD4 T cells was proposed. Naturally, these studies helped me to be skeptical of Janeway's and Matzinger's models when first put forward.

I will briefly outline the model[20], see Figure 17, and then comment on why it appeals to me.

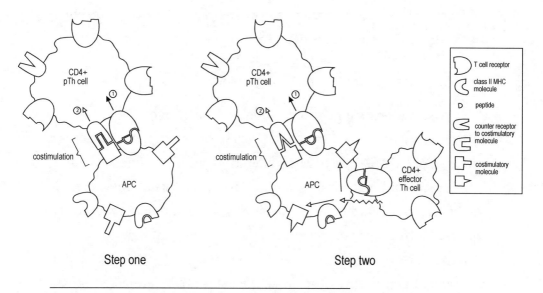

Figure 17. The Two Step, Two Signal Model of CD4 T cell activation

According to the model, the activation of a naïve CD4 T cell, specific for a nominal antigen Q, occurs in two steps. In the first, the naïve CD4 T cell interacts with Q presented by a macrophage or a mature dendritic cell. The CD4 T cell receives signal 1 and a costimulatory signal, resulting in its multiplication. These CD4 T cells become inactivated in time, probably by dying, unless these step-1 primed CD4 T cells complete step 2.

The second step involves the interaction of step 1-primed target CD4 T cells with a Q-specific B cell that presents Q. The B cell must be activated by Q-specific helper CD4 T cells to express the costimulatory molecules required to generate the costimulatory signal necessary for the step 1-primed target lymphocytes to complete step 2, and so give rise to effector CD4 T cells.

Note that the two costimulatory signals of step 1 and step 2 are most likely mediated by different costimulatory molecules and counter receptors. In fact, quite a few pairs of such costimulatory molecules and their receptors have been characterized, so this possibility is not extravagant. It is easier to imagine how step 2 can be obligatory if the two costimulatory signals are different.

Considerations for and against the model

Many observations support the model. In particular, it is consistent with our studies suggesting CD4 T cell cooperation facilitates the activation of CD4 T cells. Several different types of costimulatory molecules are present on activated, but not on naive, B cells. Thus, there is precedence for the inducible expression of the costimulatory molecule expressed by the B cell involved in step 2.

As we have seen, B cells can act as antigen-specific APC. The requirement that the APC be an antigen-specific B cell in step 2 was proposed for two reasons. First, it explains our observations showing that the CD4 T cell collaboration required to facilitate the activation of CD4 T cells involves the operational recognition of linked antigens. Our studies show that the activation of Q-specific target CD4 T cells can be facilitated by R-specific helper CD4 T cells in the presence of Q-R. In this case, the B cell could have antibody receptors specific for either Q or for R. Both types of B cell would take up the conjugate Q-R and present peptides derived from Q and R. Such B cells could mediate CD4 T cell cooperation by a mechanism involving the operational recognition of linked epitopes. Second, if linked recognition was not required, a helper CD4 T cell specific for a foreign antigen F could, in the presence of F, help the activation of a step 1-primed target CD4 cell specific for a peripheral self antigen, thus interfering with its inactivation. The requirement for the operational recognition of linked epitopes allows the ***Principle of Non-Interference*** to be realized.

A single CD4 T cell will only complete step 1 and so its progeny will be inactivated, whereas its activation through both steps requires other lymphocytes specific for the antigen. Thus, this model provides an explanation for peripheral self-nonself discrimination consistent with the one lymphocyte/multiple lymphocyte model for the inactivation/ activation of lymphocytes. This explanation of peripheral tolerance at the level of CD4 T cells is also consistent with the Historical Postulate. Note that a condition of this mechanism being realized is that a single CD4 T cell must not yield so many progeny, following step 1, that they can mutually help themselves through step 2.

More recent and beautiful in vivo imaging studies by Marc Jenkins and colleagues trace the fate of single CD4 T cells as they are activated. These imaging studies appear to show that a naive CD4 T cell first interacts in the lymph node with a dendritic cell or macrophage, divides, and then, at about day three, migrates to an area where it can interact with an antigen presenting B cell. These findings are consistent with the envisaged two step process.

Finally, many studies over several years try to address whether B cells are required to fully activate naïve CD4 T cells. The conclusions drawn are discordant. There is a recent discussion in the literature between Cohn and myself on what to do when facing such a dilemma. [2, 21, 22] Personally, I never anticipate complete concordance of ideas and observations, as some of the many observations and assumptions are likely to be wrong.

Considerations concerning the PAMP/Danger Models

In evaluating a conceptual framework, it is important to contrast its plausibility with alternative frameworks. In addition, there often are attractive features of opposing frameworks. It is important to consider whether the attractive features of one framework can be compatible, or better still even incorporated, into an opposing framework. It is most important to understand the sense in which alternative frameworks are incompatible.

History has shown that some passionate disputes should never have taken place. Historians of immunology know of the great fight, in the late 1800s between the Germans, who believed immunity was mediated by antibodies, and the French, who believed it was mediated by cells. These were not logically incompatible alternatives, and so there was no necessity, other than emotions, for a dispute. Therefore, for reasons of clarity and rigor, I first consider in what sense the PAMP/Danger Models are significantly incompatible with the Two Step, Two Signal Model that I favor.

The PAMP/Danger Models propose that whether antigen activates or inactivates CD4 T cells critically depends upon whether PAMPs or danger are present or absent at the time antigen impacts upon the immune system. The Two Step, Two Signal Model postulates that whether antigen activates or inactivates CD4 T cells depends critically upon whether lymphocyte cooperation, in the form of B cell-mediated CD4 T cell collaboration, takes place. These are incompatible proposals. To clarify the significance of this incompatibility, I first consider some consequences of the PAMP/Danger Models that make them somewhat implausible to me as understood in their strict formulation. I then consider some features I like, and their relationship to the Two Step, Two Signal Model.

I formulated the *Principle of Non-Interference* as I tried to find a way of articulating my concerns about the implications of the PAMP/Danger Models. It seems likely that PAMPs could activate APC beyond those that are presenting the antigens of the pathogen, including APC presenting peripheral self antigens. Similarly, the Danger Model, as I understand the concept, could lead to activation of APC presenting both foreign and peripheral self antigens, if the pertinent APC exist at a dangerous site. I feel it likely that such imprecise and non-specific control over whether CD4 T cells are activated or inactivated, if it had been adopted by nature, would too readily result in autoreactivity and autoimmunity. Also, I have only been able to envisage how a relatively reliable mechanism of peripheral tolerance can be achieved if it conforms with the Historical Postulate. The PAMP/Danger models are not consistent with it. They propose that whether antigen activates or inactivates a naïve CD4 T cell at a particular time depends only on the present circumstances. According to the Historical Postulate, whether antigen activates or inactivates a lymphocyte depends upon whether the immune system was continuously exposed to the antigen in the past. A major consideration in the formulation of the Two Step, Two Signal Model was that a viable model should conform to the Historical Postulate.

Last, an impetus for Matzinger's Danger Theory was that immune responses could be generated against skin grafts between, for example, mice belonging to different strains. As these foreign grafts do not express PAMPs, the Janeway formulation seemed to be at least incomplete. There is no doubt that skin grafting must provoke innate stress responses on the part of the injured recipient, responses that could be interpreted as a sign of danger. I must admit, though, to my puzzlement at Matzinger's response to her concern about Janeway's proposal. I too had the same reservations and thoughts. However, not just foreign grafts can be rejected. Many complex, foreign, non-PAMP expressing vertebrate antigens generate immune responses when administered to mice. A classical example is sterile sheep erythrocytes injected intravenously, with a very sharp needle. Where is the danger in this case?

I consider one experiment to be both elegant and inconsistent with PAMP/Danger Models. Janeway made his PAMP proposal in 1989. In 1992, he and his colleagues found a means of generating in mice CD4 T cells specific for the self-antigen, mouse cytochrome c. They found that if they immunized mice with cytochrome c, given in CFA, they could not induce these cytochrome c-specific CD4 T cells. However, they found they could do so in mice given cytochrome c in exactly the same way if these mice were in addition injected with activated mouse cytochrome *c*-specific B cells. These observations seem, at face value, to be incompatible with PAMP/Danger Models. If mouse cytochrome *c*-specific CD4 T cells exist, they should be activated upon immunizing with this antigen in CFA. I admit that reading these observations reported in 1992 gave me fortitude in publishing in 1999 the Two Step, Two Signal Model. Obviously, it seemed plausible that giving mice activated cytochrome *c*-specific B cells allows step one primed CD4 T cells to more readily complete step 2. The observations reported by Janeway and colleagues are striking.

One feature of the PAMP/Danger Models is attractive to me. It seems natural that the immune system exploits the interaction of PAMPs with PRRs and other danger signals to regulate the immune response. When I envision the situation where an individual has a systemic infection, resulting in some interactions between PAMPs and PRRs of cells belonging to internal organs, it makes eminent sense that these signals of alarm have been exploited by evolution to increase the size and intensity of immune responses. However, this is a different possibility from the one in which PAMPs or danger determine whether an immune response is generated in the first place.

Chapter 9

A theory of immune class regulation

We have seen that there are two major arms of immunity, the cell-mediated and humoral arms. We have also acknowledged that this useful distinction is an oversimplification. There are four classes of antibody in people, namely the IgA, IgE, IgM and IgG classes, and four subclasses of IgG antibody, IgG_{1-4}. Moreover, these different classes/subclasses are differentially regulated. We know this must be biologically significant as different classes of immunity, and the different subclasses of IgG antibody, bring different innate mechanisms of attack to bear upon the invader. This clearly represents quite a sophisticated system, and it is a something of a daunting task to understand its basis.

We start by considering how cell-mediated and humoral responses might be differentially regulated. In the older literature, and even today, there is little emphasis on experimentally determining which of the IgG subclasses of antibody are predominantly produced under given circumstances.

Cell-mediated and humoral immune deviation

We have seen that the generation of an antibody response to an antigen can preclude a subsequent attempt to induce DTH, against the same antigen, by a challenge that results in a DTH response in a naïve animal, see Figure 7 of Chapter 4. We refer to such a state as humoral immune deviation. Other studies revealed the existence of a state of cell-mediated immune deviation.

In the mid-1960s Mitchison found that the repeated administration of low doses of antigen to immunocompetent mice over a period of several weeks could render them unresponsive for the production of antibody upon a challenge that results in antibody production in naïve mice, as shown in point 2 versus point 1 of Figure 18. The repeated administration of higher doses of antigen primed the mice for a secondary antibody response, as seen in point 3 versus point 1. Mitchison suggested that this experimentally induced unresponsive state corresponded to a state similar to the unresponsiveness towards self antigens that we refer to as tolerance. He called this unresponsiveness *low-zone paralysis* to include a reference to the low dose of antigen needed to generate the unresponsiveness.

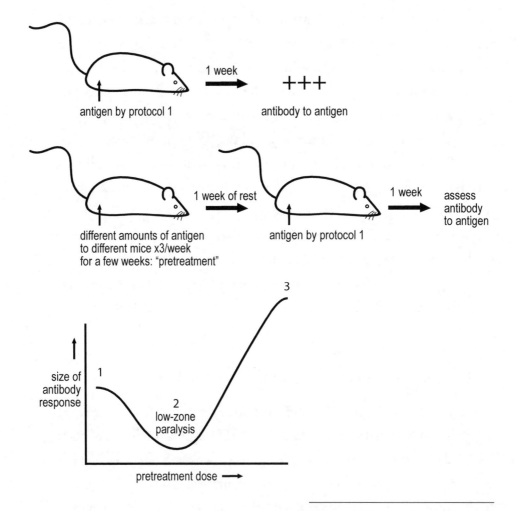

Figure 18. Induction of low-zone paralysis

In the late 1960s, Chris Parish did a similar series of experiments in rats. He employed a different antigen, but the general scheme of his experiments was similar to Mitchison's. A critical difference was that Parish measured the state of DTH to the antigen at the time of the challenge. He found that the rats unresponsive for antibody production expressed DTH against the antigen, and those primed to produce a secondary antibody response, did not, see Figure 19. He suggested that this unresponsiveness did not reflect a state of self-tolerance but one of *cell-mediated immune deviation*. Moreover, Parish's findings fit with others. For example, Salvin had shown in the 1950s, as we have seen, that a dose of antigen too low to stimulate the production of antibody generates an exclusive DTH response (Figure 6).

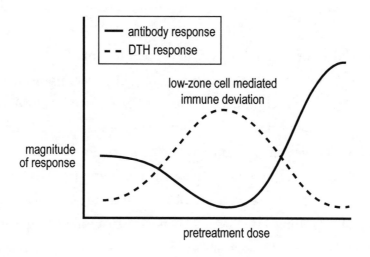

Figure 19. Parish's demonstration of low-zone cell-mediated immune deviation

Most immunologists over the years accepted Mitchison's interpretation of his observations, though it seemed obvious to Chris Parish and to me that self tolerance must reflect unresponsiveness for the generation of both cell-mediated and antibody responses. I have to say that the lack of discussion and will to resolve this issue among immunologists is surprising to me. I will suggest later that determining which alternative is more plausible might have resulted in immunologists developing effective vaccination against AIDS and tuberculosis. It is a fact of today that many immunologists are too young to know of these observations or, if older, these observations have disappeared from their horizon. This assessment represents a mini-sociological analysis on my part. I have deliberately asked young and older immunologists about their knowledge of these studies and the related phenomena.

Formulation of a theory of immune class regulation

I became interested in the early 1970s in speculating on what the physiological significance of immune class regulation might be and how the differential regulation of distinct classes might be achieved. Moreover, I anticipated and hoped that these two subjects might be related in a meaningful way. Such a connection seemed likely, given that there are distinct classes of immunity, that these are differentially regulated, and that the nature of the weapons that are brought to bear upon an invader is determined by the class of immunity the invader provokes. Moreover, as we have seen, the class of immunity generated is often critical to the efficacy with which the immune system contains an invader.[23]

It is often difficult to convey in words why a multifaceted theory is attractive. This is because words tend to result in linear, logical thought. Often understanding a theory can be likened to appreciating a piece of sculpture. You need to walk around it to realize why it is miraculous. Giving one description of a theory in words is like seeing a sculpture from just one vantage point. So, to describe a multifaceted theory well, it is helpful to describe it in linear words a number of times from different perspectives.

The cellular basis of humoral and cell-mediated immune deviation

A major question concerning immune class regulation is the nature of the mechanism responsible for the exclusivity between the generation of cell-mediated and antibody responses. Asherson's and Parish's observations respectively illustrate that states of humoral and cell-mediated immune deviation can exist.

Two reports in the early 1970s described how it was possible to generate an unresponsive state for the production of antibody to the antigen sheep red blood cells (SRBC). Peter McCullagh in Australia established one system in rats, and Richard Gershon's laboratory in the US established a similar system in mice. The systems shared two remarkable and novel features. In both, the unresponsiveness was associated with T cells that could specifically inhibit the antibody response to SRBC. The spleen of such unresponsive rats contains T cells that can inhibit the anti-SRBC antibody response when given to a recipient rat challenged with a dose of SRBC that would otherwise generate a potent antibody response. These cells at the time were called suppressor T cells, a practice I shall follow.

Second, and initially somewhat paradoxically, such unresponsive rats and mice were actually primed to make large anti-SRBC antibody responses. This became clear when both labs found ways to break the unresponsive state, and an anti-SRBC antibody response occurred that was much more rapid and of much greater intensity than a normal primary antibody response. It appeared that the antibody response to SRBC had been primed and that, at the same time, these suppressor T cells had been generated to hold this response in check. The lid on a garbage can contain flies that are multiplying inside, but removing its lid and jiggling the can allows the flies to reveal themselves.

Both Gershon and McCullagh interpreted their unresponsive states as reflecting the unresponsiveness the body naturally displays against self antigens. They both proposed that anti-self immune responses are held in check by suppressor T cells specific for self-antigens. This was a much discussed and popular view for several decades, and some still favor such models today.

I initially thought, on reading these reports, that a different view was worth considering, for three reasons.[23, 24] First, Gershon's and McCullagh's interpretation was difficult to reconcile with our Two Signal Model of lymphocyte activation/inactivation and the explanation it provided for how peripheral self-nonself discrimination is achieved. Second, it seemed highly unlikely that an unresponsive state, corresponding to self-tolerance, would have the highly primed state for antibody production characteristic of both systems. Third, such priming might be expected if this unresponsive state had a different physiological significance, as I shall now consider.

Salvin had shown that an immune response often goes through an exclusive cell-mediated phase before antibody is produced. Given that there is a high premium for the rapidity of immune responses in combating multiplying pathogens, it seemed natural that, during the exclusive cell-mediated phase, clonal expansion of those lymphocytes required for antibody production takes place. I therefore thought it plausible that the antibody response might well be primed during the exclusive cell-mediated phase of an immune response, and could be held in check at the same time by suppressor T cells. This mechanism might also account for a state a cell-mediated immune deviation following the chronic stimulation of a cell-mediated response.

I proposed in my Theory of Immune Class Regulation that Peter McCullagh's and Richard Gershon's observations reflected states of cell-mediated immune deviation. I postulated that the suppressor T cells they characterized were generated under conditions leading to an exclusive cell-mediated response, and that these T cells suppressed the production of antibody as shown. These suppressor T cells were envisaged to be central to the mechanism that inhibits antibody production during the course of the exclusive cell-mediated phase of immune responses.

Salvin's observations also showed that, as antibodies are produced, the expression of cell-mediated immunity, in the form of DTH, usually decays. I proposed that this decay was associated with the generation of T cells that inhibit cell-mediated responses, in the form of DTH. There was then no direct or indirect evidence for such cells, but the virtue of the proposal I made was that it led to testable predictions. These proposals are summarized in Figure 20.

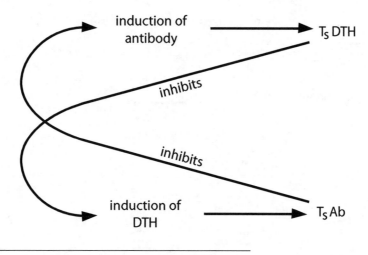

Figure 20. Proposal for how suppressor T cells ensure exclusivity of cell-mediated and antibody responses

I formulated this theory when still in Mel Cohn's lab. Circumstances led me to go to Canberra shortly thereafter to the microbiology department of the Australian National University, to which Chris belonged. When I arrived in Canberra, a graduate student of Chris Parish's, Ian Ramshaw, became interested in the Theory of Immune Class Regulation.[23] Ian and I decided to design experiments together to test the proposals outlined in Figure 20. It was quite a struggle to find appropriate experimental systems. Once we achieved this, we were amazed and excited by the cleanliness of the observations Ian made. He further characterized the surface antigens on the two types of suppressor T cells envisaged, leading to the conclusions shown in Figure 21.

It also seemed likely, as the generation of an antibody response was associated with CD4 T cells able to help antibody and suppress DTH responses, that these two activities might be mediated by one CD4 T cell. This turns out to be the case, as we shall see in the next chapter. I refer to this as a CD4 ThAb/TsDTH T cell. To summarize our take on the significance of Ian's findings, we felt the demonstration of the existence of TsAb and TsDTH cells allowed us to understand how cell-mediated and antibody responses tend to be exclusive.

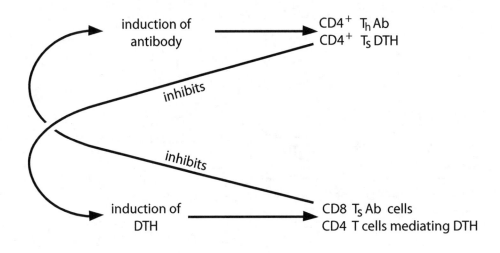

Figure 21. Summary of conclusions testing the role of suppressor T cells in ensuring exclusivity of cell-mediated and antibody responses

Distinct CD4 T cell subsets

Ian also looked at the surface antigens of the T cells able to mediate DTH, which he showed were CD4 T cells. Our observations thus defined two types of CD4 T cells. One subset of CD4 T cells suppresses DTH and helps antibody responses. The TsDTH/ThAb subset of CD4

T cells is generated under conditions that results in antibody production, and the cells belonging to this subset do not mediate DTH. Cells
of the other CD4 T cell subset are generated under conditions that
result in DTH responses, mediate DTH, and do not suppress DTH
responses. I think Ian's findings were the first evidence for distinct CD4
T cell subsets. We shall describe in the next chapter the properties and
significance of these and other CD4 T cell subsets.

The significance of the variables of immunization that affect the cell-mediated/ humoral nature of the immune response

Ian's experiments, demonstrating the means of achieving exclusivity,
left open the question of why some circumstances of immunization
favor the generation of cell-mediated immunity in the first place, and
others, of antibody responses. This question was also addressed by
the theory.[23]

Consider the hypothetical situation where we know how antigen
interacts differently with the cells of the immune system to result in
cell-mediated and antibody responses. We collectively refer to these
processes as the *decision criterion* controlling the cell-mediated/
humoral nature of the immune response. Knowing the nature of the
decision criterion would surely allow us to understand why different
variables of immunization favor the cell-mediated/humoral nature of
the ensuing immune response.

We can consider the reverse situation. When knowing the different
circumstances that favor the generation of cell-mediated and antibody
responses, we can explore what hypothetical decision criteria can
account for these circumstances. This is the approach I took in the early
1970s when trying to examine what might be the nature of the decision
criterion. I now describe the variables of immunization then known to
affect the cell-mediated/humoral nature of the immune response. This

description will provide a context for delineating the considerations that led to a proposal as to the nature of the decision criterion.

i) Dose of antigen

We have described Salvin's studies of the 1950s, see Figure 6. Low doses of antigen induce a cell-mediated response and higher doses an antibody response. The major point is that similar observations have been made in different species of animal and with diverse antigens. Salvin employed protein antigens in guinea pigs. George Mackenass and colleagues showed in the 1960s that there were similar dependencies of the class of immunity on antigen dose in the response of mice to the much more complex antigen SRBC administered by different routes. We made similar observations in the 1990s for the slowly growing intracellular parasite *Leishmania major* on infection of mice, as we shall discuss later, as well as for the response of mice to mycobacteria, the genus to which the bacterial pathogens responsible for tuberculosis and leprosy belong. Bryce Buddle and colleagues showed that the number of mycobacteria employed for infection similarly affects the cell-mediated/antibody nature of the ensuing immune response in cattle.

ii) Time after antigen immunization or after pathogen infection

This parameter was also studied by Salvin. The immune response often goes through a cell-mediated phase before antibody is produced. This too seems not to be an incidental finding but to be true under diverse circumstances. An interesting example we shall discuss several times is after infection with HIV. An exclusive cell-mediated response is generated shortly after infection. Later, when anti-HIV antibody has been produced, the individual is said to have **seroconverted**. Most interestingly, infected individuals only get really ill sometime after they have seroconverted, usually passing through a defined series of pathological stages if untreated. The time before serious AIDS symptoms are present is referred to as the honeymoon period.

iii) Previous exposure to antigen/invader

We have seen that past exposure to an antigen can dramatically alter the class of immunity generated upon re-exposure. In the most extreme cases, pre-exposure can result in immune deviation of different types. It is worth noting how important a phenomenon immune deviation may be for medical purposes. For example, based on the existence of the honeymoon period after HIV-1 infection and on other grounds, I will argue that ensuring a stable cell-mediated response to HIV-1 after infection would likely constitute effective vaccination. This perspective has focused my mind on exploring how cell-mediated immune deviation might be achieved, as I discuss in the last two chapters.

iv) The nature of the antigen

In the late 1960s, Pearson and Raffel published a paper addressing how the nature of an antigen itself can influence the kind of immunity generated. Based on their observations and other reports in the literature, they proposed that certain antigens could only generate a cell-mediated response. They identified these antigens as being either small, or larger but predominantly identical to a self-antigen. They pointed out that these antigens share the property of being minimally foreign. They also made the point that if such a minimally foreign antigen was coupled to a more foreign antigen, itself immunogenic for an antibody response, immunization with the conjugate could result in the production of antibody to the minimally foreign antigen. There were therefore B cells specific for the minimally foreign antigen. This generalization struck me forcefully. I refer to this proposal as the ***Pearson/Raffel generalization***.

The physiological significance of the existence of different classes of immunity and their differential regulation

Circumstances leading to the generation of autoreactivity

We have seen in Chapter 4 that rabbits, rendered neonatally unresponsive to BSA, are unresponsive to a challenge of BSA at three months of age, but will respond at this age to a challenge of HSA with the production of antibody, some of which binds to BSA, see Figure 8. We refer to this situation by saying that a challenge antigen that crossreacts with an antigen against which an unresponsive state has been induced can break the unresponsive state. We shall shortly argue that this type of situation reflects natural conditions under which autoimmunity can be induced.

It is helpful to be aware of some history to achieve a valid perspective. In the early 1970s, when I was thinking about immune class regulation, no one talked of central tolerance. There was no evidence for it, whereas there was indirect evidence for a mechanism of peripheral tolerance. The considerations I shall shortly outline were formulated in the early 1970s, and the considerations, though still valid, appeared more forceful in the context of the then-current belief that the immune system, to achieve tolerance, solely relies on what we now refer to as peripheral tolerance.

Weigle, whose classic 1961 paper on the BSA/HSA system we have just alluded to, subsequently explored conditions under which real autoimmunity could be induced. There is an autoimmune disease called Hashimoto's Disease, after its discoverer, in which antibodies and T cells are found against antigens prevalently if not uniquely expressed in the thyroid gland. One autoantibody regularly found is against thyroglobulin, a prominent protein made by this gland. Weigle defined in rabbits circumstances under which such antibody could not be induced, and others under which it could be. Immunization of rabbits with rabbit thyroglobulin was unsuccessful, but immunization with turkey thyroglobulin resulted in the production of anti-rabbit thyroglobulin antibody. The unresponsiveness to rabbit thyroglobulin was broken by immunization with a crossreacting antigen, chicken thyroglobulin. From a modern perspective, such a process of breaking the unresponsive state is only pertinent to understanding how autoimmunity can be induced against peripheral self antigens.

A paradoxical situation

A major reason for espousing the Two Signal Model of lymphocyte activation was its ability to account for peripheral tolerance. This model allowed us to understand why the impingement of a foreign antigen, which crossreacts with a peripheral self antigen, can result in autoreactivity against this peripheral self antigen. One aspect of this scheme disturbed me. Infections by intracellular pathogens, including viruses, are a common occurrence. An infected cell inevitably crossreacts with its uninfected counterpart. Would this not often result in the activation of anti-self lymphocytes, if present? Moreover, there were reports that, with sufficiently sensitive techniques, autoantibodies can be regularly detected to some self antigens in most healthy people. The paradox I perceived is this: our Two Signal Model was designed to explain how peripheral tolerance occurs. One might hope it would also account for some circumstances under which autoimmunity could be induced, and it did. However, it would seem that the activation of autoreactive lymphocytes would frequently occur upon infection by intracellular pathogens. So, our explanation of self-tolerance seemed to invoke a not-too-robust mechanism. This was indicated by the prevalence in healthy people of low levels of autoantibodies to several self-antigens. How could such autoantibodies be tolerated?

Although I think the concern in this line of reasoning is still valid, it lost some of its urgency once it became apparent, some years later, that central tolerance is the major means of preventing autoreactivity. The considerations just elaborated upon are still pertinent in the context of peripheral tolerance.

The relationship between the conditions required to generate different classes of immunity and their effector functions

A major impetus for my theory came from reading a study by Humphrey and Dourmashkin of the 1960s on the requirements for obtaining the IgG antibody-dependent, complement-mediated lysis of red blood cells (RBC). In these studies, the complement was present in excess, so lysis of the RBC would occur if the IgG antibody bound to the red blood cells in a manner that initiated the complement cascade. These researchers investigated how the amount of RBC lysis depended on the amount of IgG anti-RBC antibody present. They interpreted their results as showing that at least two IgG molecules had to bind to neighboring sites on the red blood cell to initiate the complement cascade. They estimated that several hundred thousand IgG molecules had to bind to the surface of a RBC for there to be a 50% chance of the RBC being lysed. I felt this conclusion to be astounding, and it led to three thoughts.

The first and inevitable conclusion was that IgG antibody-dependent complement-mediated lysis would not occur if a cell bore fewer sites, recognized by antibody, than a certain high number. Presumably, minimally foreign cells could not be effectively held in check by such an IgG antibody-dependent mechanism. In this case, it would only make sense if minimally foreign cells only induced cell-mediated immunity, as Pearson and Raffel had already proposed, and if such minimally foreign cells were susceptible to cell-mediated attack in the form, for example, of cytotoxic T lymphocytes (CTL). I add here a subsequent finding. Certain white cells have receptors for the constant parts of antibody molecules. When a foreign target cell is covered with antibody, these white cells can lyse the target cell subsequent to binding to the many antibody molecules on the target cell. Investigators have determined the number of antibodies needed on the target cell to result in lysis. It is around a million antibody molecules!

Second, this inference that IgG antibody would be ineffective against minimally foreign cells fitted in with a number of other observations. It was consistent with the observation that foreign grafts, between MHC identical strains and bearing minor histocompatibility antigens, were rejected by cell-mediated but rarely by antibody dependent mechanisms. Moreover, such grafts rarely stimulate the production of antibody. Animal studies during the 1960s on immune responses to tumors were interpreted as showing that cell-mediated responses were required to contain tumors, whereas progressive tumor growth was often found to be associated with antibody responses. It made sense that tumor cells were usually not very foreign. Third, these studies made me realize that low levels of autoreactive IgG antibody might usually be benign, and not have pathological consequences.

As I pursued this line of thought, I concluded that much of what was then known about immune class regulation could be accounted for as serving physiological needs in terms of three rules.

The first rule was that minimally foreign cells or antigens would only induce a cell-mediated response, since only that could be effective. This rule was observationally supported by the Pearson/Raffel generalization. Both antibody and cell-mediated responses could in principle be effective against more foreign antigens and cells. However, given the stringent antigen recognition requirements for the IgG antibody dependent activation of effector mechanisms, it seemed likely that low levels of IgG antibody would be ineffective. Only low levels of antibody could be produced shortly after infection, before there was sufficient time for clonal expansion of lymphocytes to the level required to produce effective levels of antibody. I thought a second rule would be that an effective response should be made as rapidly as possible that can deal with the invader. The implication of this rule is that a cell-mediated response should first be made shortly after infection, as found by Salvin (Figure 6). The downside of this proposal is that the generation of any activated autoreactive, cell-mediated lymphocytes would be more damaging than the corresponding IgG antibody response.

Recognizing this consequence led to the third rule. The immune response should not be more vicious than necessary to satisfy the second rule, as this would only result in unnecessary autoimmune damage. Thus, once sufficient clonal expansion has taken place so that effective levels of IgG antibody can be produced, the immune response evolves into a humoral mode, as is known to occur and shown in Figure 6. This evolution minimizes the damaging consequences of any autoreactivity induced and yet allows an effective response against the invader.

These three rules constitute a scheme according to which there is a meaningful relationship between the nature of the effector functions mediated by a class or subclass of immunity and the conditions under which this class/subclass is generated. The rules provide a view as to why it is evolutionarily advantageous that distinct immune classes exist and are differentially regulated in the manner observed. Such regulation optimizes the generation of effective immune responses whilst minimizing the damaging consequences of any autoreactivity generated. We shall see later that autoreactivity does not always lead to damaging autoimmunity. In many cases, it appears the damaging autoimmunity is due to a cell-mediated response, whereas comparable humoral autoreactivity can be relatively benign. This conclusion has implications for strategies of treating patients with autoimmune conditions.

A proposed decision criterion: The Threshold Hypothesis

Most of the variables of immunization affecting the cell-mediated/humoral nature of the ensuing response have quantitative aspects. These include the dose of antigen, the time after antigen impact/infection at which the nature of the immune response is assessed, and the quality of the antigen, by which I mean its degree of foreignness. Perhaps the most interesting question I faced when thinking about the nature of the decision criterion was how the Pearson/Raffel generalization, that minimally foreign antigens can only generate a cell-mediated response, could be realized in cellular terms. This generalization is so central because of its relationship to the envisaged physiological significance of immune class regulation, as just outlined.

How could the degree of foreignness of an antigen be assessed? This is clearly not an intrinsic property of an antigen, but reflects a relationship between it and the host. Rat red blood cells are self antigens in rats and not very foreign in mice, as rats and mice are both rodents and rather closely related in the evolutionary sense. However, rat red blood cells are very foreign from the perspective of chickens. From what we know, there are very few mature lymphocytes specific for self antigens, only a few specific for minimally foreign antigens, and more for very foreign antigens. It seemed to me that an assessment of the foreignness of an antigen must reflect the fact that there are fewer lymphocytes specific for minimally foreign antigens than for more foreign antigens. Since I believed CD4 T cell cooperation is likely required to activate CD4 T cells, and such activation is an early event in an immune response, it was natural for me to consider that it was the number of CD4 T cells that was important. I therefore proposed that tentative antigen-mediated CD4 T cell cooperation favored a cell-mediated response, in the form of DTH-mediating CD4 T cells, CD8 CTL and TsAb cells, and robust antigen-mediated CD4 T cell cooperation favored an antibody response in the form of CD4 ThAb/TsDTH cells.

I refer to the idea that different thresholds of CD4 T cell cooperation are needed to generate cell-mediated and antibody responses, in other words tentative and robust CD4 T cell cooperation respectively, as the *Threshold Hypothesis*, which is illustrated in Figure 22. This Figure, adapted from papers describing the hypothesis when first proposed, reflects the fact that it was made a good decade before it was realized that the T cell recognizes antigen in an MHC-restricted fashion. This depiction therefore reflects the assumption that T cells recognize intact antigen similarly as antibody. I examine later a formulation that takes account of the fact that T cells recognize antigen in an MHC-restricted manner.

My students and I have thoroughly tested predictions of this hypothesis in diverse systems over the last forty years. It seems to me that there is little reason to doubt its essential correctness[24, 25] Moreover, as I also explain below, it accounts for all the variables of immunization know to affect the cell-mediated/humoral nature of the immune response.

Implications and evidence for the Threshold Hypothesis

Consider first a situation a few days after impact of a foreign antigen upon the immune system or after an infection, when an exclusive DTH, cell-mediated response is evident. One of the earliest events after antigen impact is the multiplication and differentiation of CD4 T helper cells. In this case, so long as the level of antigen is sustained, CD4 T cell collaboration will get stronger with time, and so the response is likely to evolve from a cell-mediated into an antibody mode. This evolution is in accord with Salvin's observations, as seen in Figure 6.

Now consider the situation where an antibody response is generated in mice at day seven after administration of the antigen. If we imagine a similar situation, except we immunize the mice with a lower antigen dose, the degree of the antigen-mediated CD4 T cell collaboration will be reduced. If we reduce the antigen dose sufficiently, the CD4 T cell collaboration will no longer be robust but, if sufficiently reduced, will become tentative, so a cell-mediated response will be generated in place of the antibody response. These considerations are again in accord with Salvin's findings on how the cell-mediated/humoral nature of the response depends upon antigen dose.

The Threshold Hypothesis thus accounts for why minimally foreign antigens can only induce cell-mediated immunity; for how the nature of the immunity depends, for more foreign antigens, on the dose of antigen administered; and for how the immune response evolves with time after antigen impact or after infection. One can make a unique prediction based on the Threshold Hypothesis. Consider a situation in which an antibody response is generated. If we keep all variables constant, except we find a way to reduce the number of CD4 T cells present, the response will, at some point, change from an antibody to a cell-mediated mode.

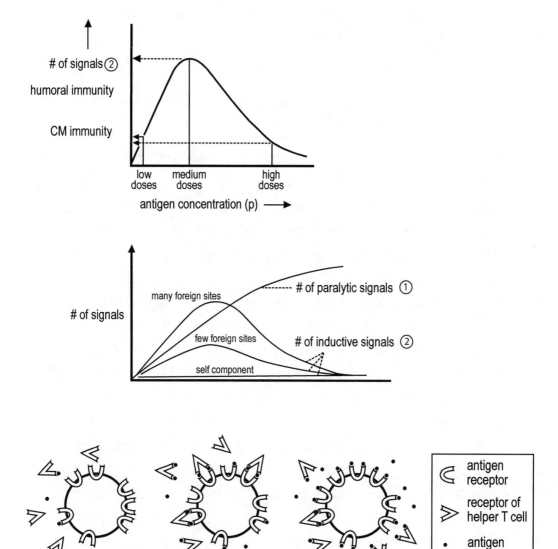

Figure 22. The Threshold Hypothesis

Evidence

My students and I have tested predictions of the Threshold Hypothesis extensively in diverse systems over the last forty years.[24, 25] I feel our evidence is strong and convincing, but we have been unable to engage the immunological community to give the hypothesis or the evidence significant consideration. I would like to illustrate three important conclusions drawn from our experiments.

The cell-mediated/antibody nature of the immune response depends on the number of CD4 T cells present

It is possible to generate immune responses in vitro by culturing spleen cells with antigen. It is also possible to lethally irradiate mice and reconstitute them with syngeneic spleen cells, and challenge the reconstituted mice with antigen. We found conditions in both these systems where antigen could induce unprimed spleen cells to generate primary cell-mediated responses under some circumstances and primary antibody responses under other circumstance. We were thus able in both systems to assess what cell types, and how many cells of different cell types, are required to generate primary cell mediated and primary antibody responses. We found in both systems that more spleen cells were required to generate antibody than cell-mediated responses. Moreover, the cell required in greater numbers was identified as a CD4 T cell.[24, 25]

Thus, in a number of situations we kept all the variables the same, such an antigen dose, and time after antigen impact at which we assessed the nature of the immune response. We could change the nature of the immune response from an antibody to a cell-mediated mode by reducing the number of CD4 T cells present in an animal, or in a tissue culture well. I have recently reviewed these diverse experiments.[24, 25] Two other sets of findings are important and consistent with the Threshold Hypothesis.

The cell-mediated/humoral phenotype of the immune response depends conjointly on the CD4 T cell number and antigen dose

The cell-mediated/humoral nature of an immune response depends on both the amount of antigen and the number of CD4 T cells, in an interdependent fashion, in a manner consistent with the Threshold Hypothesis. For example, fewer CD4 T cells are required to generate either a cell-mediated or an antibody response with a medium dose of antigen than with a low dose. This is expected as the CD4 T cell collaboration, required to generate T_{DTH} cells and CD4 ThAb cells, is antigen mediated.[23, 24]

The critical CD4 T cell interaction involves the recognition of linked epitopes

The third important finding is with respect to how the CD4 T cell cooperation takes place. We showed in a number of ways that the CD4 T cells could only cooperate, in a manner to determine the cell-mediated/humoral nature of the ensuing immune response, if they recognized the same nominal antigen. For example, consider two non-crossreactive antigens R and Q. Consider also the situation where there are a constant number of CD4 T cells specific for the nominal antigen Q. The cell-mediated/humoral nature of the anti-Q immune response is affected by the dose of Q and is not affected by the presence of CD4 T cells specific for the nominal antigen R, even when R is present. On the other hand, if we replace the antigen Q with the conjugate Q-R, then the R-specific CD4 T cells do affect the cell-mediated/humoral nature of the response to Q, in a manner anticipated within the context of the Threshold Hypothesis. The cooperation of the CD4 T cells, whose strength determines the cell-mediated/humoral nature of the response, is mediated by the recognition of linked epitopes by the interacting CD4 T cells.[23, 24, 25] How might this be achieved?

How can the operational recognition of linked epitopes be achieved?

We have already seen that the CD4 T cell interaction, required to activate naïve CD4 T cells, requires the recognition of linked antigenic epitopes, and that such linkage is probably achieved as a consequence of an antigen-specific B cell mediating this cooperation between CD4 T cells. We suggest the determination of whether CD4 T_{DTH} or CD4 ThAb/TsDTH cells are primarily generated occurs in step two of the Two Step, Two Signal Model for the activation of CD4 T cells. We consider this mechanism of CD4 T cell cooperation, involving the recognition of linked epitopes, to be physiologically most significant, for reasons we now address.

We know that a sufficiently strong cell-mediated T_{DTH} response can contain *Mycobacterium tuberculosis*, the pathogen responsible for tuberculosis, and that a mixed cell-mediated/humoral or predominant humoral response leads to chronic or progressive disease. Infection by nematodes, a type of worm, normally results in an effective antibody response. We have also seen that a strong antibody response to an antigen is associated with CD4 T cells that can suppress a cell-mediated response to this same antigen. We have evidence that the TsDTH cells, generated during an antibody response and able to inhibit the generation of DTH, act via the recognition of linked epitopes, as outlined in *Rediscovering*. Consider what would happen if this were not the case. In an individual infected simultaneously with *M tuberculosis* and a nematode, the ThAb/TsDTH cells specific for the nematode would, in the presence of nematode antigens, suppress the protective cell-mediated, T_{DTH} response against *M tuberculosis*, thereby undermining the generation of this protective response. Given the importance of the cell-mediated/humoral nature of an immune response against an invader, we envisaged that the decision criteria, determining the cell-mediated/humoral nature of immune responses to two concurrent infections, would be independent. I refer to this proposal as the ***Principle of Independence***. When assessing whether this principle is true, we found conditions under which we could generate a DTH response in mice to the antigen Q following intravenous immunization with a low dose of Q, and other conditions where we could generate a predominant antibody response to the non-crossreacting antigen R by intravenous immunization with a high dose of R. We found that mice, immunized with both antigens in a manner similar to the singly immunized mice, delivered by a mixture of the antigens from the same syringe, mount indistinguishable responses as singly immunized mice, see Figure 23. This and related findings support the idea that the determination of the cell mediated/humoral nature of the immune response is consistent with the Principle of Independence. Our proposed mechanism for the decision criterion explains how this principle is realized, as different antigen-specific B cells mediate the CD4 T cell collaboration determining the cell-mediated/humoral nature of these immune responses generated against different, non-crossreacting antigens.

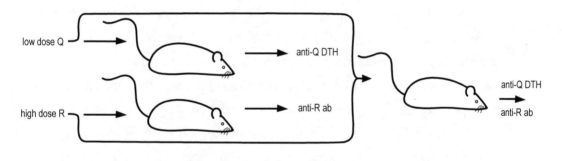

Figure 23. Observations testing the Principle of Independence

Chapter 10

Further aspects of immune class regulation

We shall make a case in the next and final chapter that an understanding of the nature of the decision criterion, controlling the cell-mediated/humoral phenotype of an immune response, is central to the rational design of interventions in diverse areas of medicine. In preparation for these considerations, we address three topics in this chapter.

We consider the basis of the major proposals for the nature of the decision criterion controlling the cell-mediated/humoral nature of the immune response currently entertained by the immunological community. Second, we consider the plausibility of these proposals. Third, we take a broader but brief overview of immune class regulation in an attempt to assess the limits and the strengths of the knowledge we now have and how it might fit into a more comprehensive picture. However, before embarking on this narrative, it is useful to describe one technical development that has had a great impact upon the field.

The generation and characteristics of T cell clones

Studies in the mid- to late 1970s showed that primed T cells, stimulated with antigen, could produce soluble molecules that aided the multiplication and differentiation of B cells into antibody-producing cells. These early studies were taken to indicate that T helper cells could, when appropriately stimulated, produce antigen non-specific molecules that affect the growth and differentiation of target B cells. Their production by T cells was envisaged to partially represent their helper function.

Primed T cells can similarly produce, on stimulation with antigen, molecules that act on target T cells. The characterization and cloning of the gene for one such molecule had a major impact on the field, as I shall shortly relate.

Such molecules produced by lymphocytes that act on and thereby affect the differentiation state of other cells are called *lymphokines*. Molecules made by one white blood cell that influences the differentiation state of other white cells are referred to as an *interleukins*, a messenger between white cells (leukos, Greek for white). There are quite a number of different kinds of white cells, including lymphocytes and different kinds of APC. A more general term for these molecules is *cytokine*, a name with no implications on the cell type that produces it or the target cell on which it acts. An *autocrine* cytokine, interleukin, or lymphokine acts on a target cell similar in type to the cell that produces it. A *paracrine* cytokine, interleukin, or lymphokine acts on a target cell that is of a different kind than the cell that produces it. All three can have both autocrine and paracrine activities. One of the most important such molecules is *interleukin-2 (IL-2)*.

The cytokine IL-2 is produced by some CD4 T helper cells. Its biological importance is largely due to its autocrine and paracrine abilities in causing the proliferation of all partially activated T cells, including those CD4 T cells that make IL-2 and others that do not. These biological properties of IL-2 have been exploited to good effect to achieve technical ends.

A major difficulty in studying antigen-specific lymphocytes is their natural scarcity. Immunologists discovered that they could put a few primed T cells in a culture well with antigen, together with IL-2 and a source of APC free of other T cells, and about a week or two later, antigen-specific T cell colonies specific for the antigen were found in some wells. If a few primed T cells are initially placed in a well, it could be shown that these colonies are most often derived from a single T cell. These colonies are referred to as **T cell clones**. Coffman and Mosmann showed that the CD4 T clones they derived were of two basic types, distinguishable by the cytokines they produce upon antigen stimulation. Th1 cells typically produce IL-2 and IFN-γ, and Th2 cells produce IL-4 and IL-10. The significance of these cytokines will become apparent in the coming sections. As a first rule of thumb, the generation of Th1 cells is associated with DTH responses and of Th2 cells with antibody responses.

Decision criteria controlling the Th1/Th2 phenotype of the immune response

The Cytokine Milieu Hypothesis

A paper in 1993 showed that infection of macrophages with dead *Listeria monocytogenes,* bacteria well known for being responsible for a form of food poisoning, results in the production of a molecule called IL-12 by the infected macrophages. Moreover, it was shown that the presence of IL-12 caused ovalbumin (OVA) to activate OVA-specific naive CD4 T cells to differentiate exclusively into Th1 cells. This seminal paper gave rise, in the context of further observations, to the popular idea that the ambient cytokine environment, in which antigen activates naïve CD4 T cells, determines the Th1/Th2 nature of the effector CD4 T cells generated. This general proposition will be referred to as the **Cytokine Milieu Hypothesis**. We shall shortly outline what these further observations are. However, I think it helpful to point out at the beginning a potential difficulty with this idea.

Macrophages do not produce such substantial IL-12 upon phagocytosing just any antigen, or on endocytosing antigen. It is most probable that the dead *L monocytogenes* express some PAMPs recognized by the PRR of the macrophages, and that this interaction results in the observed production of IL-12. In this experimental system, the Th1/Th2 phenotype of the OVA-specific effector CD4 T cells is assessed.

However, the CD4 T cells that recognize the nominal antigen OVA most likely do not substantially crossreact with the antigens of *L mono-cytogenes*. Thus, this envisaged mechanism is inconsistent with the Principle of Independence we have already discussed. This major consideration made me question the plausibility of the mechanism when it was initially proposed.

Following this initial and influential study, it was shown employing in vitro systems that the presence of IFN-γ promotes the generation of Th1 cells and of IL-4 of Th2 cells. Such in vitro studies were in time supported by in vivo studies in a particularly enlightening animal model of a human infection that is uniquely contained by a cell-mediated, Th1 response.

Observations in the mouse model of human cutaneous leishmaniasis

The human infectious disease of cutaneous leishmaniasis, caused by the protozoan parasite *Leishmania major*, is initiated when an infected sandfly takes a blood meal and leaves parasite- containing saliva in place of the blood. In humans, the presence of the parasite can be largely limited to the site of the bite and, with the generation of an appropriate immune response, the initial lesion resolves. This course is associated with a predominant Th1, DTH response and is naturally referred to as resistance. When infection results in time in an immune response with a substantial Th2 component, the infection is not contained at the original site of infection, and the parasite spreads to distal sites, resulting in distal lesions. This process is sometimes referred to as metastasis. Such a situation reflects susceptibility.

James Howard, who established the mouse model, found that, when a million *L major* parasites are injected into mice belonging to different inbred strains, the type of immune response generated can be different, as well as the outcomes of infection, in a manner that mirrors what follows infection of humans. Mice of resistant strains mount a sustained Th1, cell-mediated response, the multiplication of the parasites is rapidly contained within about four weeks after infection, and the parasites are largely limited to the original site of infection. Mice of susceptible strains first generate a transient cell-mediated, Th1 response, but this response evolves to have a significant and often predominant Th2 component, parasite load increases, and the parasites metastasize to distal sites. This experimental model has been well characterized and exploited to understand how the Th1/Th2 phenotype of immune responses against a real infection is determined.

This model system was employed to explore the idea that the presence of IFN-γ is required for the generation of Th1 cells and the presence of IL-4 is required for the generation of Th2 responses. In fact, dramatic results are seen following the administration to resistant mice, at the time of infection, of an antibody that "neutralizes" the activity of IFN-γ. This antibody binds to IFN-γ so that the IFN-γ can no longer bind to cellular receptors for IFN-γ. These resistant mice generate a Th2 response against the parasite in place of the Th1 response and so are susceptible. In addition, the administration to susceptible mice of antibody that neutralizes IL-4 at the time of infection results in a Th1 response and so resistance. Such in vivo observations leave little doubt that sustained Th1 and Th2 responses are respectively associated with and responsible for resistance and susceptibility, and that IFN-γ is required to generate stable Th1 responses, and IL-4 to generate stable Th2 responses. We shall later discuss alternative ideas as to the significance of these findings.

The PAMP Hypothesis

We have already surmised that PAMPs associated with *L monocytogenes* stimulate macrophages to produce IL-12, a cytokine that can, under certain circumstances, favor the generation of Th1, as opposed to Th2 cells. Indeed, we can draw two conclusions from this observation. First, PAMPs can, under some circumstances, affect the Th1/Th2 nature of ensuing immune responses. Second, the PAMP and Cytokine Milieu Hypotheses are not necessarily exclusive. Some have suggested that

PAMPs are critical in determining the Th1/Th2 nature of the immune response mounted against the PAMP-containing pathogen. The question this proposal raises is not whether there can be such a role for PAMPs under some particular circumstances, but how important is this role in practice? I refer to the view that PAMPs are the primary factor in determining the Th1/Th2 phenotype of the immune response as the *PAMP Hypothesis*.

Pertinent to assessing this hypothesis is the fact that PAMPs can greatly affect the expression of different costimulatory molecules by APC. This finding has led to the proposal that PAMPs might determine the Th1/Th2 phenotype of immune responses by affecting the expression of different costimulatory molecules by APC.

The Antigen Presenting Cell Hypothesis

Many studies have delineated that there are different types of APC. These studies reflect not only distinctions already made between B cells, dendritic cells (DC), and macrophages. For example, various studies have delineated distinct subsets of DC. These subsets differ, for example, in the surface antigens they express. A rather widely held view is that antigen presentation by different subsets of DC leads to the generation of different subsets of CD4 T cells. Indeed, there is evidence that the nature of the DC can be critical, as we shall outline in the last section of this chapter. In this case, the involvement of the DC is in supporting the generation of a subset of CD4 T cells other than Th1 and Th2 cells. I consider the experimental case for a central role of different DC in critically favoring the generation of Th1 or Th2 cells to be weak. I also think there are considerations making this possibility unlikely, as outlined in the next section.

The plausibility/implausibility of the Cytokine Milieu, the PAMP and the APC Hypotheses

I suggest that there are both conceptual and observational grounds against the idea that these hypotheses are of general validity. However, this does not mean that these hypotheses cannot be pertinent in certain limited situations. Some of these circumstances will be outlined in the last section of this chapter. There are also unusual, likely pathological, conditions under which the Cytokine Milieu Hypothesis has some validity. It is useful to describe one such situation for its own interest and also to provide context.

There are cases where overwhelming parasite infections are associated with unusually large immune responses against the parasite, and the presence of high levels of parasite antigen in blood. These responses can lead to a level of cytokines in the blood, produced by parasite specific lymphocytes, beyond what can be normally detected. The presence of elevated levels of cytokines can affect the type of immune response mounted against non-parasite antigens, as observations show. However, these are exceptional circumstances. I argue below that such observations do not support the pertinence of the Cytokine Milieu Hypothesis under less extreme and more physiological conditions.

I shall first outline experimental evidence against the idea that the PAMP Hypothesis, the Cytokine Milieu Hypothesis, or the APC Hypothesis is pertinent to how the Th1/Th2 phenotype of immune responses is generally determined. I then consider the conceptual reasons that lead me to suggest they are implausible on biological grounds.

Evidence

I think the most compelling evidence against these hypotheses is our diverse observations, and those of others. They demonstrate that the Th1/Th2 phenotype of a primary immune response against an antigen can be modulated from a Th2 or mixed Th1/Th2 to a Th1 mode by changing the number of antigen-specific CD4 T cells present, with all other variables of immunization kept constant.[24, 25] The change in the number of CD4 T cells does not change the PAMPs expressed by the antigen/pathogen, nor is it expected to change the type of APCs that presents the antigen, nor to change the ambient cytokines present at the sites where naïve CD4 T cells are activated. It is a straightforward expectation based on these hypotheses that reducing the number of CD4 T cells should change the size of the CD4 T cell response, not its Th1/Th2 phenotype. A most interesting example of a situation where reducing the number of CD4 T cells changes the phenotype of the immune response, from a Th2 to a Th1 mode, is the immune response of mice susceptible to a standard challenge of a million *L major* parasites.[24, 25] We shall return to this finding in the next and last chapter on medical interventions.

It is not apparent in terms of these hypotheses how the evolution of the Th1/Th2 phenotype of the immune response, with time after antigen impact or after infection by slowly growing microorganisms or parasites, can be explained. Neither is it clear on the three hypotheses how the antigen dose- dependence of the Th1/Th2 phenotype of the immune response can be accounted for.

Moreover, this dependence of Th1/Th2 phenotype on both dose of antigen or number of organisms employed for infection, and on the time after antigen impact or infection, seems to hold true for foreign, immunogenic vertebrate antigens not expected to express PAMPs, and for pathogens such as mycobacteria and leishmania parasites that express diverse PAMPs. In view of this, it seems likely that the same mechanism operates to determine the Th1/Th2 phenotype of responses to PAMP-expressing and PAMP non-expressing entities. The threshold mechanism we envisage accounts for all these observations, as outlined in the previous chapter. I suggest in the next chapter that these considerations, on the nature of the decision criteria, are critical to attempts to develop strategies to vaccinate against pathogens uniquely susceptible to cell-mediated attack.

Conceptual considerations

The PAMP Hypothesis

This hypothesis states that the genes of the pathogen that define the PAMPs the pathogen expresses determine the Th1/Th2 phenotype of the host's immune response. From an epidemiological point of view, pathogen evolution would probably favor the generation of chronic infections with poor immune control of pathogen multiplication, thus leading to optimal infectious spread of the pathogen through the host population. This seems an unlikely scenario from the host's perspective.

The Cytokine Milieu Hypothesis

We have seen from studies in the *Leishmania major* mouse model that some cytokines are central in determining the Th1/Th2 phenotype of the effector Th cells generated following primary infection. We now recall two different ideas on how IL-4 facilitates, or is even required, to generate Th2 cells. We then look at some evidence supporting the idea we favor before considering whether we can draw general conclusions concerning the role of IL-4 and of other cytokines.

A considerable number of studies have explored, in the context of the Cytokine Milieu Hypothesis, what might be the initial source of the IL-4 envisaged to be necessary to generate Th2 cells. For example, some studies have led to the suggestion that mast cells that produce IL-4 on degranulation might be the cell stimulated to produce the IL-4 deemed important in generating Th2 cells. Other studies suggest other cellular sources of IL-4, as I have reviewed elsewhere.[24, 25]

A different possibility for the role of IL-4 comes from a knowledge of its biological properties. IL-4 is an autocrine cytokine in that it is made by Th2 cells and preferentially stimulates a population of Th2 cells to divide. In a recent study, where we had established conditions under which antigen could activate naïve CD4 T cells to give rise to IL-4-producing Th2 cells, we showed that adding antibody that neutralizes IL-4 prevented the generation of Th2 cells and resulted in the generation of Th1 cells instead. This reflects earlier observations of others but in our experimental system. In these cultures, we incubate the antigen with a purified population of CD4 T cells and a population of T cell-depleted spleen cells as a source of APC. We found that if the source of the APC was from the spleen of a mouse unable to make IL-4 for genetic reasons, we could still generate IL-4 producing Th2 cells. Thus, the IL-4 required for the generation of Th2 cells is made by the Th cells themselves. This is perhaps not so surprising in view of IL-4's autocrine nature. This and related findings have led us to articulate a different hypothesis from the Cytokine Milieu Hypothesis. We briefly outline this hypothesis and comment on ideas as to its general biological significance.

The Cytokine Implementation Hypothesis

I propose that there is usually no significant level of ambient cytokine during the first stage of CD4 T cell activation, and that the Th1/Th2 phenotype of the activated CD4 T cells generated in the initial phase is determined by the threshold mechanism outlined in the last chapter. However, parasites are complex, with many different antigenic components present in different amounts. We suppose that the threshold mechanism does not result in all naïve CD4 T cells specific for these different antigenic components of a complex antigen to be activated to generate either only Th1 cells or only Th2 cells, but to typically generate a predominant Th1 or predominant Th2 response in this initial phase. However, long-term immune responses against such complex antigens are often, though not universally, remarkably uniform in their Th1/Th2 phenotype. I refer to this as reflecting the *coherence* of immune responses. This seems to be an important physiological characteristic of the response. Thus, we propose that it is physiologically significant that the different cytokines made by Th1 and by Th2 cells have the properties of being self-promoting for the generation of the subset of CD4 T cell that produces them.

The physiological significance of the coherence of immune responses

Mycobacterium tuberculosis, which can cause tuberculosis, infects and lives inside macrophages. It is known that IFN-γ is produced by Th1 cells and delivered to the surface of infected macrophages, when effector Th1 cells recognize mycobacterial antigens presented by the infected macrophage. This delivery of IFN-γ leads to the activation of metabolic pathways within the macrophage that kill or inhibit the multiplication of the mycobacteria. Thus, IFN-γ is known to be essential to the containment of these mycobacteria by a Th1 response. However, it turns out that if mycobacterium-specific Th2 cells are present, they deliver IL-4 to the infected macrophage, whereupon the IL-4 binds to surface receptors of the macrophage causing the down-regulation of at least some of the metabolic pathways up-regulated by IFN-γ. Thus, not only is there a tendency for the generation of Th1 and Th2 cells to be exclusive, but effector Th1 and Th2 cells can have counteractive effects. It is not only important in containing mycobacteria to generate sufficient *M tuberculosis*-specific Th1 cells, but also not to generate corresponding Th2 cells that, if present, counteract the effectiveness of Th1 cells. This scenario makes plausible the physiological importance of generating coherent responses in which the large majority of CD4 T cells activated against a pathogen belong to the same subset, as often observed.

The role of Th1 and Th2 cytokines in promoting coherence

I would like to consider here how the roles of the cytokines made by Th1 and Th2 cells might lead to coherence. I shall also consider how the presence of anti-IL4 neutralizing antibody could render an initial generation of CD4 T cells, that is Th2-biased and generated in accord with the threshold mechanism, to subsequently evolve to give rise to the predominant generation of Th1 cells.

Consider a situation where the initial activation of CD4 T cells by the threshold mechanism gives rise to three Th2 cells that produce IL-4 and IL-10 for every one Th1 cell that produces IFN-γ and IL-2. This response is thus initially biased in favor of the generation of Th2 cells. Most of the cytokines produced by cells of one CD4 T subset either directly favor the further generation of effector CD4 cells belonging to their CD4 T cell subset, or act to inhibit the generation of effector CD4 T cells of the opposing subset or subsets. Thus, the IFN-γ produced by Th1 cells is known to inhibit the proliferation of Th2 but not of Th1 cells, and the IL-10 produced by Th2 cells inhibits the production of IFN-γ by Th1 cells. The IL-2 produced by Th1 cells can stimulate, by itself, the proliferation of both Th1 and Th2 cells. Given all these activities, it would appear possible to understand how this initial phase of the response, resulting in three Th2 cells for every Th1 cell, could develop into a predominant Th2 mode. The IL-10 produced by the Th2 cells would inhibit the production of IFN-γ by the Th1 cells, lifting the inhibition on the proliferation of the Th2 cells so they can become dominant by multiplying in response to both IL-2 and IL-4.

In the presence of anti-IL-4 neutralizing antibody, however, IL-4 will be unable to stimulate the proliferation of Th2 cells; IL-2 has the potential to stimulate the proliferation of both Th1 and Th2 cells, but the proliferation of the Th2 cells can be held in check by IFN-γ, made by either CD4 Th1 cells themselves or the associated CD8 TsAb cells. We can thus see in this system how the presence of anti-IL-4 antibody could modulate the response from an initial bias towards the generation of Th2 cells into a Th1 mode. We suggest this particular mechanism likely reflects general mechanisms by which many CD4 T cell responses, initially somewhat heterogeneous in their Th phenotype, may become more coherent due to the self-promoting qualities of the cytokines produced by the CD4 T cells belonging to the most prevalent subset.

The DC hypothesis

This hypothesis, as currently formulated, is curiously incomplete at the conceptual level. It begs the question of why certain pathogens or antigens, which typically generate different types of immune response under given circumstances, are differentially taken up by different DC subsets under these different circumstances. This deficiency becomes even more serious when one appreciates that the Th1/Th2 phenotype of an immune response often evolves with time and when one considers the antigen-dose dependence of this phenotype.

Further aspects of immune class regulation

The strengths and weaknesses of the Th1/ Th2 paradigm of immune class regulation

Strengths

Various people, most notably Robert Coffman and Tim Mosmann, cloned CD4 T cells from in vitro propagated lines of CD4 T cells. Coffman and Mosmann characterized the cytokines produced by different CD4 T cell clones upon antigen stimulation. They found two major types of clones, referred to as Th1 and Th2. Observations support the idea that the properties of these cells reflect natural cells that are generated in vivo. For example, cells of Th1 clones produce on in vitro stimulation with antigen IFN-γ and, when transferred to a syngeneic animal to a local site with antigen, cause a typical DTH reaction. Cells harvested from mice expressing DTH to an antigen will produce IFN-γ upon in vitro stimulation with the antigen. Such studies brought a new level of molecular analysis to bear upon how immune class regulation occurs and is expressed.

Understandable limitations of the Th1/Th2 paradigm

We have already acknowledged that trying to understand immune class regulation in terms of two modes of response, the cell-mediated and humoral modes, is too simplistic to account for what we know of how different classes of immunity are differentially generated in vivo. There are more classes and subclasses of immunity than can be accounted for by such a paradigm. Correspondingly, there are limitations in trying to understand immune class regulation in terms of Th1 cells that mediate DTH and Th2 cells that help antibody responses and suppress the generation of DTH responses. We now try to understand what these limitations are and how our description can be elaborated to become more realistic.

The general conditions under which the two major recognized forms of cell-mediated immunity, DTH and CTL, are generated are similar. They are both generated before antibody is produced in large amounts and, when generated exclusive of antibody production, appear to be associated with exclusive Th1 responses. Once substantial antibody is produced, the generation of both CTL and DTH-mediating cells usually declines. Thus, these two subclasses of cell-mediated immunity appear to be roughly coordinately expressed. However, a considerably more complex picture becomes apparent when we consider humoral immunity.

I start by considering what the likely correlates are in mice and humans, in terms of antibody production, of exclusive and predominant Th1 and Th2 responses, and mixed Th1/Th2 responses. I shall then consider in the next section what is known about the regulation of the production of other classes/subclasses of antibody whose induction is not directly governed by the generation of Th1 and Th2 cells.

First, it appears possible to generate in both mice and humans exclusive Th1 responses without production of antibody. This conclusion is counter to a frequent comment in the literature that human and murine Th1 responses are associated respectively with IgG_2 and IgG_{2a} antibody production. Clinical observations in humans show that a state of DTH can be generated without production of specific antibody, and parallel observations show the same circumstance occurs in mice. However, it seems likely from observations in the literature that predominant Th1 responses, correlated with the production of a small Th2 component, are associated in mice and humans with predominant production of IgM antibody and of IgG_{2a} and IgG_2 antibody respectively.

Predominant Th2 responses seem to be associated in mice and humans with IgG_1 and IgE antibody responses. Mixed Th1/Th2 responses appear to be associated in mice with mixed IgG_{2a} and IgG_1 antibody responses and in humans probably with mixed IgG_2 and IgG_1 antibody responses.

Function and regulation of the production of other classes of antibody

Conspicuously missing from this picture are human and mouse IgA and human IgG_4 antibody responses. It is natural to look at two kinds of observations in trying to make sense of how this class and this subclass of antibody are regulated: conditions resulting in their production and the nature of their effector function.

Mouse and human IgA antibody do not facilitate the activation of any of the major mechanisms of attack that other classes of antibody do, such as complement mediated functions, phagocytosis of antibody-coated bacteria or viruses, antibody-dependent mast cell-mediated acute inflammation, or antibody-dependent cellular cytotoxicity. IgA antibody production follows sustained antigen stimulation via mucosal surfaces, and the IgA produced is delivered to these surfaces. It has been reported that individuals genetically deficient in the production of IgA antibody are more susceptible to systemic infection in giardiasis, commonly known as beaver fever. This infection is caused by drinking water contaminated with the pathogen carried and excreted by beavers. The lack of IgA antibody allows the pathogen to more readily break the intestinal barrier. IgA seems to primarily function by neutralizing pathogens and toxins. However, simple IgA deficiency does not appear to have dramatic sequelae.

Somewhat similarly, human IgG_4 antibody is not known to facilitate the activation of any known effector mechanism. Moreover, human IgG_4 has some most unusual chemical properties that have significant physiological consequences.

Observations reported about ten years ago on the structure of human IgG_4 antibody make one think radically about the role of this subclass of antibody. Similar to all other IgG molecules, IgG_4 antibody molecules are divalent for the antigen that stimulated their production when secreted by an antibody-producing cell. Most remarkably, secreted IgG_4 molecules present in blood are labile under normal physiological conditions; they split into half molecules and then recombine, with other half molecules generated from other IgG_4 molecules, to result in a hybrid molecule that is univalent for two different antigens! Such monovalent recombinant IgG_4 molecules can block the formation of antibody/antigen aggregates involving other classes of antibody. A fascinating example might have clinical implications.

Studies have examined the kind of immunity present in individuals living in a geographical area where allergies to a particular antigen are common. Allergic individuals, as expected, have predominant IgE antibodies specific for the antigen, and often high levels of IgG_1 antibodies as well. Non-allergic individuals are also immune, usually having higher levels of IgA and IgG_4 antibodies, and relatively lower levels of IgE and IgG_1 antibodies than allergic individuals do. There are sustained reports over the years that the allergic state of individuals can be temporarily relieved by giving them serum obtained from non-allergic individuals living in the same geographical area. It seems likely that IgG_4 antibody of the non-allergic individual can temporarily block the ability of the antigen to cause IgE-dependent mast cell degranulation by binding to the antigen in an overwhelming fashion.

The findings on the correlation of subclasses of IgG antibody produced in allergic and non-allergic individuals living in the same geographical area have led to plausible suggestions for treatment. Exposure of an allergic individual to the allergen in a manner that modulates the antibody response from an IgE/IgG$_1$ to an IgA/IgG$_4$ mode should constitute effective treatment. We shall return to this possibility in the next chapter.

Induction of human IgA and IgG$_4$ responses

The existence of Treg cells

A subset of CD4 T cells of more recent discovery and characterization are regulatory CD4 T cells, alternatively and more succinctly referred to as T$_{reg}$ cells. It is recognized that there are two types, distinct in terms of their origin and their specificity repertoire. Some T$_{reg}$ cells, called *natural Treg cells*, are generated and appear to be activated in the thymus by some peripheral self-antigens early in the life of a young mouse. This is remarkable in that it had been assumed, before these findings were reported that, on interacting with self-antigens in the thymus, all T cells would be obliterated by the type of mechanism proposed by Lederberg and discussed in Chapter 2. The activation of natural T$_{reg}$ cells in the thymus involves presentation of some peripheral self antigens by special APC. Most interestingly, treatment of a mouse so they are relatively deficient in natural T$_{reg}$ cells results in diverse forms of autoimmunity.

The other major type of T$_{reg}$ cells, referred to as *induced Treg cells*, is generated upon sustained antigen stimulation at mucosal surfaces. This is a manner of antigen stimulation known for many more years to result in the production of IgA antibody. Thus, sustained consumption of antigen will usually result in IgA antibody being produced against the antigen and often in an inability to generate other classes of immunity upon immunization by non-mucosal routes. Why might the generation of induced T$_{reg}$ cells and production of IgA both follow sustained antigen stimulation via mucosal surfaces? It turns out that both natural and induced T$_{reg}$ cells characteristically produce the lymphokines IL-10 and TGF-β. It also turns out that IL-10 enhances in people the production of IgG$_4$ antibody and TGF-β the production in mice and people of

IgA antibody. Moreover, it appears that T_{reg} cells can inhibit the antigen-dependent production of cytokines by Th1 and Th2 cells in an antigen specific manner. Thus, cells of this CD4 T cell subset likely display, through the cytokines they produce, some of the dominating and self-promoting characteristics of the cells of the Th1 and Th2 subsets.

The IL-10 producing subset of CD4 Treg cells

A beautiful longitudinal study was carried out over a few successive years on the state of immunity of beekeepers to bee venom antigens. These individuals have an allergic reaction every spring on receiving the first few stings of the year. After a couple of weeks, corresponding to about the first fifty stings of spring, the IgE-mediated inflammatory reaction to stings has disappeared. This disappearance is associated with the appearance of another subset of CD4 T cells that produce IL-10 but not TGF-β. Most interestingly, the appearance of this subset is associated with increased production of IgG_4 but not of IgA antibody levels. Bee stings of course involve sensitization by antigen exposure through the skin, a non-mucosal route.

Variants of the Hygiene Hypothesis

It appears that the prevalence of allergies, and allergic dependent asthma, is increasing in industrialized countries. A proposal was made some years ago, primarily on epidemiological grounds, that this increase is due to a lower incidence of exposure of young children to microorganisms and parasites in industrialized and more hygienic societies. This proposal was given some support by the evidence that even individuals living in industrialized societies have a lower incidence of allergies if living in what is deemed a less hygienic environment. For example, children living on family farms have a lower incidence of allergies, a life style suggested to be less hygienic than that of their urban counterparts. This proposal, called the *Hygiene Hypothesis*, was formulated to explain the increased incidence of allergies in industrialized societies.

The original hypothesis also proposed an immunological mechanism by which hygienic conditions could result in increased allergies. This proposal was made at a time when Th1 and Th2 were the only recognized subsets of CD4 T cells. It was suggested that in less hygienic environments, exposure to various microorganisms and parasites led to Th1 responses and consequently biased subsequent immune responses towards Th1 and away from Th2 modes. It was anticipated that, in the relative absence of such exposure, Th2 responses associated with IgE production and allergies would naturally be more readily generated.

This proposal now seems unlikely, as we shall shortly discuss. However, the idea that the degree of exposure of the young to microorganisms and parasites can affect the way the adult immune system responds to new challenges is potentially important, plausible, and intriguing.

One fact and another consideration led to a realization that the hypothesis in its original formulation was somewhat implausible. First, not only is there an increased incidence of allergies in industrialized societies, but also of autoimmunity, most often expressed as cell-mediated, Th1 autoimmunity. It is found in a number of cases that autoreactivity can be expressed either as damaging Th1 autoimmunity, or as less damaging or benign Th2 autoreactivity. It thus seems that a hygienic environment favors not only an increased incidence of Th2-dependent allergies but of Th1 autoimmunity. These generalizations are difficult to reconcile with the Hygiene Hypothesis as originally conceived.

Moreover, it was never really plausible that exposure to microorganisms and parasites early in life should uniquely bias the immune system towards the subsequent generation of Th1 responses. Such exposure would surely often result in Th2 responses, and priming for them. We shall later review the basic idea of the Hygiene Hypothesis in the context of more recent findings on the existence and role of T_{reg} cells.

The decision criteria controlling whether antigen generates Th1, Th2, or induced Treg responses

The rather speculative view developed here is based upon three major and tentative generalizations.

First, the specificity of natural T_{reg} cells is towards some peripheral self-antigens expressed in an appropriate manner in the thymus to cause their differentiation into this subset. Induced T_{reg} cells have a specificity repertoire reflecting the requirements to generate them: chronic stimulation, usually at mucosal surfaces by foreign antigens such as gut flora (intestines) or allergens (lung, intestines). In addition, a number of observations lead to the generalization that chronic stimulation with substantial amounts of foreign antigen at mucosal surfaces is required to produce IgA and IgG_4 antibody. It thus appears that natural and induced T_{reg} cells are specific for different types of antigen.

Second, the nature of primary immune responses to antigens delivered to the intestine, assessed shortly after antigen impact or infection, depends upon antigen dose or the number of infective organisms.[24, 25] Low doses or numbers favor Th1 responses, while higher doses or numbers favor Th2 responses. It appears that roughly the same rules govern acute immune responses caused by sensitization at mucosal surfaces as at non-mucosal sites. However, it has been known for a long time that chronic immunization at mucosal surfaces with substantial amounts of antigen leads to production of IgA and IgG_4 antibody.

Third, when people do not respond anaphylactically to increasing doses of an allergen, administered as part of *specific immunotherapy (SIT)* to treat an allergic state, the nature of their immunity to the allergen changes. This therapy results in a deviation of the response away from an IgE and IgG_1 towards an IgA and IgG_4 mode. Similarly, when children naturally grow out of an allergic state to milk products, it seems likely that there is a similar modulation in the nature of the immune response. In summary, just as immune responses can evolve from a Th1 to Th2 mode, so it appears that mucosal immune responses can evolve from a Th2 to a Th3/T_{reg} mode, the alternative Th3 designation being employed to indicate their role in facilitating IgA and IgG_4 antibody responses. One possible way of explaining these patterns is to include the generation of Th3 cells in the Threshold Hypothesis. More prolonged and greater antigen stimulation is required at mucosal surfaces to generate Th3 than Th2 cells. This possibility, to be plausible, must account for why antigen stimulation by mucosal surfaces is required for the production of IgA but not for IgG_4 antibody.

Roles for PAMPS and APC in regulating the generation of activated CD4 T cells belonging to other CD4 T cell subsets than the Th1 and Th2 subsets

The role of PAMPS

The subset of CD4+ Th17 cells characteristically produces IL-17. Studies show that the substantial generation in mice of this subset can depend upon whether the guts of the mice are colonized by certain segmented bacteria. It seems likely that these bacteria express PAMPs whose presence facilitates the generation of Th17 cells.

Retinoic acid produced by intestinal APC

Observations show that dendritic cells, derived from intestinal tissues, can uniquely support the in vitro generation of $Th3/T_{reg}$ cells under certain circumstances. One feature of intestinal DC that distinguishes them from other, non-mucosal DC is their production of retinoic acid. Interestingly, non-mucosal DC can support the generation of T_{reg} cells under these same circumstances if supplemented with retinoic acid.

The role of retinoic acid in this context is interesting. Its production by these intestinal APC seems to be constitutive. It is different in this sense from the production of IL-12 produced when *Listeria monocytogenes* infects macrophages. As discussed above, the envisaged mandatory role of PAMP-induced IL-12 in promoting Th1 responses seems unlikely in view of the desirability, and evidence for, independence in the determination of the Th1/Th2 phenotype of immune responses. However, the potential role envisaged for retinoic acid is different. As its production is not primarily triggered by antigen, but is constitutive, it can be seen to bias all chronic immune responses to a mode in which T_{reg}/Th3 cells and IgA and IgG_4 antibody are produced. This envisaged mode of action does not violate the Principle of Independence.

Finally, it seems plausible that natural and less hygienic conditions than those obtaining in industrialized societies, especially early in life, results in greater exposure to microorganisms and parasites, and favors the long-term generation of Th3 cells at the expense of Th1 and Th2 cells. This variation of the Hygiene Hypothesis accounts for the increased incidence in industrialized societies of both Th1 autoimmunity and Th2 dependent allergies. It can also explain why the ablation of natural T_{reg} cells can result in Th1 autoimmunity.

Chapter 11

Medical interventions

Some general considerations

An envisaged central role for B cells in mediating T cell/T cell interactions

We have seen that the antigen-mediated B cell/Th cell collaboration required to produce antibody, the antigen-mediated CD4 T cell/CD4 T cell collaboration involved in the activation of CD4 T cells, and the antigen-mediated inhibition of the generation of DTH by T_sDTH CD4 T cells, all involve the operational recognition of linked epitopes. This feature is critical, as only in this case can the regulation by these T cells of immune responses be specific, resulting in such features as *non-interference* and *independence*, as already discussed. I suggest that the action of all antigen-specific T cells in facilitating or inhibiting the activation of other lymphocytes involves a mechanism that results in the operational recognition of linked epitopes. I further suggest that, given the MHC-restricted nature of antigen recognition by CD4 and CD8 T cells and the fact that these T cells do not share the critical functional characteristics of APC in processing and presenting exogenous antigens, that interactions between these T cells must, to achieve linked recognition, involve antigen-specific B cells as the mediating APC. These considerations and this generalization are important, if valid, in designing strategies to prevent or treat immunologically related clinical situations.

The frequency of antigen-specific CD4 T cells belonging to different subsets and the behavior of the population to which they belong

A particularly enlightening study addresses how populations of antigen-specific lymphocytes behave with respect to cytokine production upon antigen stimulation in terms of the relative numbers of antigen-specific lymphocytes belonging to different subsets of CD4 T cells present in the population. Moreover, the conclusions drawn from this study are likely pertinent for the development of strategies of prevention and of treatment.

A method was employed to isolate allergen-specific CD4 T cells belonging to different subsets of CD4 T cells, as defined by the cytokines they produce upon antigen stimulation. Such isolation was carried out with lymphocytes from different individuals living in the same geographical area, some of whom were allergic and others of whom were non-allergic to the antigen. Three types of antigen specific CD4 T cells were isolated and enumerated: those that characteristically produce IL-10, IL-4, or IFNγ upon antigen stimulation. It was found that cells belonging to all three types of CD4 T cell subset could be found among the *peripheral blood lymphocytes (PBL)* of both allergic and non-allergic individuals, but that their relative frequencies were different. The frequency of IL-4 producing Th2 cells is, not surprisingly, greater in allergic individuals, and of IL-10 producing Th3 cells in non-allergic individuals. Also, the cytokines made by the two populations were characteristically different upon antigen stimulation.

The cells of these three CD4 T cell subsets, isolated from a non-allergic individual, were artificially reconstituted to be at the frequencies characteristic of the PBL of allergic individuals. The reconstituted population of PBL behaved, upon antigen stimulation, as the PBL population of an allergic individual with respect to cytokine production.

This fascinating reconstruction study suggests that the relative frequencies of these three CD4 T cell subsets determine the behavior of the population as a whole with regard to their cytokine profile. There is probably not a missing, critical, and as yet undefined other cell type.

The potential of immunological imprinting

We have seen that immune responses to antigens can be locked into a cell-mediated, Th1 mode, and into a humoral, Th2, IgG_1/IgE mode. We also suggest that human responses can be locked into a T_{reg}/Th3, IgA and IgG_4 mode. I propose that our knowledge of immune class regulation can be exploited to design ways of modulating immune responses specific for a particular antigen into any desirable mode of these three, and that such an ability would have considerable medical consequences.

Suppose we purposely immunize all healthy individuals at a young age with an antigen C so the response to C is locked into a cell-mediated, Th1 mode; with an antigen H so the response to H is locked into a Th2, humoral mode; and with an antigen R so the response to R is locked into a T_{reg}/Th3 mode. The antigens C, H, and R would be chosen not to crossreact. We refer to such a healthy individual as *imprinted*. We shall shortly discuss how such imprinting might be achieved.

Suppose we also identify a clinical condition in an imprinted individual in which an immune response to one or more antigens, referred to as A, is deemed critical in causing the pathology. Suppose we also know that the clinical condition is severe when the mode of the anti-A immune response is of one type, say M1, and the condition benign when the mode is of another type, say M2. It is clear in principle how we might prevent this clinical condition from arising. We would immunize individuals with A in such a manner that we cause an M2-imprint upon the immune system.

Consider further the case where the M2 mode, associated with a benign state, is the Treg/Th3 mode. We expect to be able to facilitate the generation of a protective M2 imprint against A by immunizing with A under conditions known to generate a Th3 response and imprint. We might further facilitate this by immunizing under such conditions with the conjugate A-R. The R-specific T_{reg}/Th3 lymphocytes, and other potentially associated specific lymphocytes, should facilitate the deviation of the primary anti-A immune response into a T_{reg}/Th3 mode, and result in an A-specific T_{reg}/Th3 imprint.

Similarly, it may be possible to treat a patient making a predominant M1 anti-A response by exposing them to A-R conjugates. Note that administering A-R conjugates to a patient already immune to A will have the tendency to pull the response against A towards a T reg/Th3 mode and pull the immune response against R towards an M1 mode. We can perhaps favor the former modulation over the latter by simultaneously administering relatively moderate and larger amounts of A-R and R respectively. I anticipate such stimulation will favor the stability of the anti-R Th3 state, whilst modulating the anti-A response towards a Th3 mode.

I realize such proposals are futuristic, but I hope this description might foster their eventual exploration.

Observations show the importance of cell-mediated immune imprinting to attain medical and clinical goals
Vaccination

BALB/c mice are the prototypical strain susceptible to *Leishmania major* as, on standard challenge of a million parasites, a predominant anti-parasite Th2 response is rapidly generated, associated with both parasite metastasis to sites other than the original site of infection, and uncontrolled parasitemia. We employed this model to examine whether we could establish low-zone, cell-mediated immune deviation against the parasite. Infection with a hundred or a thousand parasites resulted in a sustained Th1 response for several weeks. We found these mice resisted a normally pathogenic challenge of a million parasites eight weeks after being infected. This resistance is associated with a polarized Th1 response and so containment of the parasite. We say these mice are

Th1-imprinted as a result of infection with low numbers of parasites, We refer to this strategy as the ***Low Dose Vaccination Strategy***.

I recognized there is a barrier to achieving universally effective vaccination. The genetically diverse constitution of individuals usually results in different types and strengths of immune responses when these various individuals are subjected to the same standard challenge. We explored whether there might be a means of vaccinating genetically diverse individuals with a standard challenge resulting in a Th1 response and Th1 imprints in all vaccinated individuals.

If the threshold mechanism really operates in diverse individuals to determine the Th1/Th2 phenotype of the immune response, it means that relatively low and high numbers of slowly growing organisms will respectively generate in the long term Th1 and Th2 responses. We experimentally tested this idea as described in *Rediscovering*. Briefly, we injected mice belonging to different strains with widely different numbers of a given strain of parasite and by the same route. We followed the ensuing parasitemia. Two major patterns were found in all strains of mice. Mice resisted a parasite challenge with a relatively low number of parasites and suffered progressive disease with a relatively high number. We showed that resistance and progressive disease were respectively associated with stable Th1 responses and, in the long term, with predominant Th2 responses. We could define a ***transition number*** of parasites for a given strain of mouse for a given route of infection. Infection with a number of parasites below the transition number resulted in a stable Th1 response, and infection with a number of parasites above the transition number resulted in responses with a significant Th2 component. Remarkably, the value of this transition number varied as much as 100,000 fold for different strains of mice! I might add that we repeated this kind of experiment with a second parasite strain and when infection is by different routes. Thus, the dependency of the Th1/Th2 phenotype of the ensuing immune response on the parasite number employed for infection seems to be a general rule.

In light of this generality, we suggest that infection of genetically diverse individuals with a number of slowly growing organisms, below the transition number for all individuals vaccinated, will result in time in a Th1 response and Th1 imprints in all individuals, and thus provide universally efficacious protection. We examined in young mice whether infection with few BCG, the attenuated mycobacterial strain used to vaccinate against human tuberculosis, could generate Th1 responses and Th1 imprints. In the older literature on BCG infection of mice, a million BCG are referred to as a low number. We found infection of young mice with twenty mycobacteria could generate the most potent Th1 responses and Th1 imprints.

The BCG vaccine has been employed in past attempts to protect cattle against tuberculosis without success. Buddle and colleagues dropped the number of BCG employed for infection a million-fold, and found they could get robust protection against a normally pathogenic challenge of **Mycobacterium bovis**, the pathogen responsible for cattle tuberculosis.

Treatment

Visceral leishmaniasis, caused by **Leishmania donovani**, is a human disease that is rapidly fatal if untreated. Infected individuals may show no symptoms, in which case they are designated as a healthy infected individual, or they may suffer progressive disease and be a patient. Healthy infected individuals express strong DTH to leishmanial antigens and produce low levels of antibody. Patients express little or no DTH to parasite antigens, but produce antibody. Patients are given a three-week treatment with a drug that kills the parasite and has toxic side effects, so the time of treatment has been minimized. Most interestingly, individuals who are successfully treated by this short course of drugs do not become ill again, even when living in an endemic area.

We looked at the prevalence of different Ig isotypes among anti-parasite specific IgG antibody. As outlined in *Rediscovering,* we found that the ratio of IgG_1 to IgG_2 antibodies was high in patients and low in both drug-cured individuals and in healthy infected. The ratio of these two isotypes reflects the Th1/Th2 phenotype of the immune response: a high ratio reflects a predominant Th2 response, a medium ratio a mixed Th1/Th2 response, and a low ratio a predominant Th1 response.

Two tentative conclusions seem warranted from these observations. The immune response can be modulated backwards from a mixed Th1/Th2 to a predominant Th1 mode upon drug treatment. This treatment is known to fail if the parasite is genetically resistant to the drug. Therefore, it seems likely that successful drug treatment results in a decrease in antigen load, resulting in turn in a modulation of the response from a Th1/Th2 to predominant Th1 mode. Thus, it appears that the amount of antigen not only affects the Th1/Th2 phenotype of primary immune responses, but the phenotype of on-going immune responses.

We established in the mouse model of cutaneous leishmaniasis a state of chronic disease associated with an immune response with a mixed Th1/Th2 phenotype. We found ways of curing this chronic state and modulating the immune response to a predominant Th1 mode. Most interestingly, a highly effective treatment was partial depletion of CD4 T cells. Thus, the number of CD4 T cells is not only important in determining the Th1/Th2 phenotype of primary immune responses, but seems to control the Th1/Th2 phenotype of on-going immune responses.

In summary, these observations lead us to suggest that antigen load and CD4 T cell number similarly control the Th1/Th2 phenotype of on-going immune responses as they do of primary responses. We shall assume the correctness of this likelihood in our consideration of strategies for the treatment of different clinical conditions to be shortly outlined.

Allergies

There is currently a standard way of attempting to treat allergies by a process called *specific immunotherapy (SIT)*. Antigen, against which the patient is allergic, is given in escalating doses in a manner that modulates the response from an IgE, IgG_1, Th2 to an IgA, IgG_4, Th3 mode. Administering the allergen can precipitate acute anaphylaxis. The administered allergen can bind to mast cell-bound IgE antibody and so cause acute inflammation. The ensuing reaction can be lethal, particularly if systemic. It is for this reason that attempts at desensitization are initiated with low amounts of antigen and signs for mild, acute inflammation monitored so treatment can be stopped if it appears that the further administration of the antigen may cause too strong a reaction. When treatment can be completed and is effective, it results in a modulation of the immune response against the allergen to that found in non-allergic individuals.

Two aspects of this specific immunotherapy are evident. First, it appears that, just as there is often a natural progression from a Th1 to Th2 mode upon continued antigen stimulation, so there can be a natural progression from a Th2 to $Th3/T_{reg}$ mode. It appears to be the basis of specific and successful immunotherapy. Second, achieving such a progressive modulation by administering the allergen is problematic, as it can cause anaphylaxis.

I describe three treatments by which immune responses in allergic patients might be more effectively modulated by administering different forms of the pertinent antigen. Each attempt exploits the fact that T cells recognize antigen in a different manner from antibody and B cells. This fact might allow one to develop a means of stimulating allergen-specific CD4 T cells with forms of antigen that are not recognized at all, or not well recognized, by the allergen-specific antibody present in the patient, thereby avoiding or minimizing anaphylaxis.

One approach is already the subject of extensive on-going research. Oligopeptides, around 10-20 amino acids in length and corresponding to stretches of the primary sequence of the polypeptide chains of the allergen, are administered to the experimental and allergic animal. The idea is that antibody will not bind the oligopeptides and, even if some minimal binding occurs, it is unlikely to be multivalent and able to trigger mast cell degranulation. This approach has had some significant success in animal models. The basis of the effects seen is currently not very clear.

We anticipate that the transition from a Th2 to Th3/T$_{reg}$ mode involves antigen-dependent CD4 T cell cooperation, mediated by antigen-specific B cells. In this case, single oligopeptides are not likely to be efficient in causing this modulation, as there will be relatively few CD4 T cells specific for a single peptide. Moreover, CD4 T cells specific for different peptides are unlikely to optimally interact during the potential modulation of the response from a Th2 to Th3 phenotype in a manner we envisage to be necessary, unless presented by the same B cell. Is there a way we could design an antigen that is anticipated to be highly immunogenic in stimulating allergen-specific CD4 T cells, but would not interact well with allergen-specific IgE antibody present in an allergic individual? If there was, we might be able to employ such an antigen to stimulate the allergen-specific CD4 T cells from a Th2 to a Treg/Th3 mode, whilst minimizing the risk of anaphylaxis.

We know that the stimulation of CD4 T cells specific for the nominal antigen A requires it to be processed into a series of peptides of the order of 10-20 amino acids long, which we will call a1, a2, a3, and so on. Some of these peptides will bind to the grooves of the host's class II MHC molecules, and so are presented and stimulate their corresponding CD4 T cells. Consider the gene for A. We can divide the gene into a series of segments that code for oligopeptides about fifteen amino acids in length. We can design a gene related to that encoding A, where the DNA segments are in a different order so that, when the encoded protein is processed, it would give rise to the same series of peptides, as well as some more. My graduate student David Kroeger suggested that we call this synthetic protein an *isopeptidogen* of A, as it generates the same peptides as A when processed. We anticipate the isopeptidogen of A will react poorly with anti-A IgE antibody, as its three dimensional structure will be very different from that of A, and so administration of the isopeptidogen may be useful for desensitizing individuals allergic to A, whilst minimizing the likelihood of anaphylactic reactions. In fact, the isopeptidogen would most likely not form a well-defined three-dimensional structure, and would be insoluble. It may therefore be possible to give it locally in a manner that it could be a source of antigen to stimulate immune cells. A further advantage of such administration would be that the antigen, being insoluble and so primarily limited to a local site, would not so readily cause systemic anaphylaxis as does systemic antigen.

Finally, this idea could be extended to develop a more elaborate strategy. This would involve immunizing the isopetidogen linked to an antigen R against which a $Th3/T_{reg}$ cell imprint had been established. We anticipate this would further facilitate the modulation of the immune response against A into a Th3 mode.

Autoimmunity and transplantation

We consider one approach to prevent and one to treat autoimmunity. The strategy of prevention is based upon the idea that CD4 T cells are central orchestrators of immune responses and that, if we are able to ablate the generation of autoreactive CD4 T cells, we would prevent autoimmunity.

Autoimmune diabetes

I illustrate this strategy in the context of autoimmune diabetes. There is a fascinating animal model for human autoimmune diabetes, the *non-obese diabetic (NOD)* autoimmune mouse strain. The incidence of autoimmune diabetes is a polygenic trait in both mice and humans, and there are remarkable parallels in the genetics governing autoimmune diabetes in the two species. This parallelism encourages the use of the NOD mouse as an appropriate model for the study of the human disease. Virtually all individuals who come down with autoimmune diabetes are known to be at risk from their family history.

One phenomenon of the autoimmune response has been carefully delineated in mice and, to a significant but lesser extent, in autoimmune people. This phenomenon is central to our proposed strategy of preventing this disease. Autoimmune diabetes is due to immune responses against the β-islet cells of the pancreas that produce insulin. Investigators have examined how the repertoire of NOD CD4 T cells, specific for the β-islet cells of the pancreas that produce insulin, changes with age. It is found that the CD4 T cells detectable at three weeks of age in NOD mice are predominantly specific for one antigen, that we shall denote as the *initiating antigen*, and the range of β-islet antigens recognized by CD4 T cells increases with age in a fairly specified and prescribed manner. We refer to these other β-islet antigens, as well as the initiating antigen, as *target antigens.* Moreover, classical studies show that an ablation of T cells specific for the initiating antigen, by injecting it into the thymus of NOD mice about a week or so old, ablates the generation of CD4 T cells specific for the initiating antigen as well as CD4 T cells specific for the other target antigens.

This increase with age in the repertoire of the β-islet specific autoimmune CD4 T cells also occurs in humans and is referred to as *epitope-spreading.* The word epitope was originally employed as a term for a structure recognized by an antibody or by a lymphocyte via its antigen specific receptor. There could be several different and non-exclusive mechanisms underlying the phenomenon of epitope spreading. It seems likely that one mechanism reflects the requirement for CD4 T cell cooperation in the antigen-dependent activation of CD4 T cells. Moreover, this explanation also explains why ablating the generation of CD4 T cells specific for the initiating antigen results in failure to generate CD4 T cells to other target antigens.

As already noted, autoimmune diabetes is a polygenic disease, but the MHC is the most important locus, and certain class II MHC alleles are virtually essential for the development of the classical disease in both mice and man. Moreover, the human and mouse MHC molecules predisposing to diabetes have idiosyncratic features in common. I take all these observations as supportive of the idea that the activation of CD4 T cells specific for one or two initiating peptides is critical to instigate the autoimmunity. Then the repertoire of anti-self CD4 T cells can increase, as reflected in the phenomenon of epitope spreading. Moreover, we know of unobtrusive means to ablate CD4 T cells specific for a given peptide by administering it systemically. I suggest it is also technically feasible to determine the nature of these initiating peptides. These considerations lead to the following proposal of how to prevent autoimmune diabetes.

As outlined above, family history and the inheritance of certain genetic alleles can be used to identify individuals at high risk for developing autoimmune diabetes. The identification of the few initiating peptides may allow one to develop preventive measures. These peptides could be systemically administered to individuals at risk at an age well before diabetes occurs in a manner known to ablate the corresponding CD4 T cells. Such administration, if early enough, may constitute an effective preventive measure. Experience will inform us whether successive treatments of administering the initiating peptides are required for long term prevention.

Studies with NOD mice also demonstrate that the establishment of Th2 cells specific for one peptide of a β-islet antigen in a mouse with predominant Th1 responses to the β-islet antigens can change the IgG subclass of antibody generated against these antigens. I suggest this finding indicates a predominant Th2 response against these diverse β-islet antigens. It seems likely that the β-islet antigens are operationally linked, thereby accounting for how epitope spreading occurs. Such linkage also allows one to understand how the generation of Th2 cells specific for one peptide of a β-islet antigen can, if generated in sufficient numbers, help to modulate the response to other β-islet antigens from a Th1 to Th2 mode. We refer to such a process as *epitope conversion.* It may be possible to treat early diabetes by immunizing with a conjugate of one or more β-islet antigens with H or R, in H and R imprinted individuals. I anticipate the response to β-islet antigens may be modulated into a harmless Th2 or Th3 mode in this case.

In addition, it is interesting to note that autoimmune diabetes is generally regarded as being caused by Th1 autoreactivity against β-islet antigens. A current treatment is to deplete the patient's B cells. This makes sense if B cells are required to generate and sustain CD4 T cell activation, including sustained Th1 responses, as envisaged in the Two Step, Two Signal Model of CD4 T cell activation.

Multiple Sclerosis

Last, we consider another related form of treatment for multiple sclerosis based upon our understanding of immune class regulation. It is believed an inflammatory, Th1, cell-mediated response directed at the insulating myelin sheath of nerve cells of the central nervous system is responsible for pathology in human multiple sclerosis. The inflammatory response causes damage to these central nerves, impairing their function.

In the chronic, relapsing form of the disease, there are severe episodes interspersed with relatively quiescent periods. One hypothetical explanation for this pattern is based on what we know about how on-going immune responses are regulated. We have seen that the dose of antigen is important not only in determining the Th1/Th2 phenotype of primary immune responses, but of on-going immune responses. We propose that in severe episodes of the chronic, relapsing form of the disease, the inflammation around the myelin sheath caused by a predominant Th1 response results in substantial release of myelin antigen. As the antigen load increases, the immune response is consequently modulated from a predominant Th1 into a mixed Th1/Th2 mode. There is thus less damage and, as this occurs, a quiescent period ensues. As a result of this quiescence, the amount of myelin antigen released decreases, and so the response is again modulated in time from a mixed Th1/Th2 to a more predominant Th1 mode. This modulation would occur following a decreased availability of antigen by the same mechanism described above when considering the drug-treament of patients with visceral leishmaniasis.

If this hypothetical explanation is valid for how the symptoms of the disease wax and wane, the ratio of the IgG_1 to IgG_2 isotypes present in IgG anti-myelin antibodies should vary accordingly. The ratio should be small during attacks, and an increase in its size should correlate with the appearance of a quiescent phase. If this prediction were verified, it would confirm the supposition that severe episodes and quiescent periods are associated with different classes of immunity to myelin antigens. In this case, it might be ethical to explore the possibility of administering myelin antigens conjugated to other antigens against which there is a strong Th2 or Treg/Th3 response in order to establish an immune state against myelin antigens associated with quiescent periods. In imprinted individuals, immunizing with myelin antigens conjugated to H or R might constitute effective treatment.

Autologous stem cell transfer

One currently employed protocol to treat some severe cases of autoimmunity is called autologous stem cell transfer. The immune system is ablated by heavy irradiation, for example, and the individual is reconstituted with autologous (self) stem cells. The patient is obviously at risk during the time it takes the immune system to regenerate from stem cells. The process of stem cell reconstitution mimics a natural one. Tadpoles can respond to foreign frog antigens. During the process of metamorphosis from a tadpole to a frog, the tadpole's lymphocytes are hormonally ablated, and new lymphocytes are regenerated from stem cells. Autologous stem cell transfer can in principle be similarly used to re-establish tolerance to self antigens or to achieve acceptance of MHC-disparate grafts. However, these are elaborate, expensive, and invasive protocols.

Successful and non-obtrusive transplantation of MHC-disparate organs presents formidable difficulties. The frequency of a recipient's

T lymphocytes against allo-MHC antigens is roughly a thousand fold higher than for foreign minor transplantation antigens, and we have seen this frequency for allo-MHC is associated with exceptionally vigorous immune responses. It seems a daunting task to apply the first type of strategy, involving the ablation of antigen-specific CD4 T cells of the recipient, as a pretreatment for a patient about to receive an organ bearing foreign MHC antigens. It might be more practical to employ autologous stem cell transfer to achieve tolerance to donor MHC antigens, or to overcome the transplantation barrier by deviating the response against the foreign donor MHC antigens, before transplantation, into a Treg/Th3 mode.

Infectious Diseases

Effective vaccination of people in the West has only been publically endorsed against pathogens that cause relatively acute disease and can be contained by a vigorous antibody response. Protection is in large measure achieved by a secondary humoral response guaranteed by vaccination. Effective vaccination against other pathogens, particularly intracellular pathogens with the potential for causing chronic disease, has not been formally accomplished. I feel I should note, however, that some people in the Middle East have been and are deliberately exposed to sandflies infected with *Leishmania major* to protect them against cutaneous leishmaniasis, a process called **leishmanization**. It seems likely that leishmanization is a folk practice that reflects a process of natural infection with low numbers of parasites, resulting in a cell-mediated imprint.

I start by considering vaccination against HIV, the virus that causes AIDS, as an illustrative example. I shall also consider whether and how it might be possible to reverse AIDS progression in the early stages of the disease, shortly after seroconversion. I illustrate further considerations in discussing the possibilities of achieving effective vaccination against and treatment of tuberculosis.

Vaccination against AIDS

Before considering what principles might guide the development of an effective strategy of vaccination against AIDS, I would like to indicate why I believe a consideration of such principles is critical. Some argue that diverse possibilities should be explored without prejudice. I have illustrated in *Rediscovering* why such a program would lead to gigantic studies needed to cover the diverse variables of immunization thought to be potentially pertinent, without a guarantee that a winning combination of variables would be found.

Consider our actual experience in realizing the low dose vaccination strategy in the mouse model of cutaneous leishmaniasis. We started with a framework in mind, as previously outlined. We injected a few groups of mice belonging to a susceptible strain with different numbers of parasites. We then followed the development, or lack thereof, of parasitemia, and the type of immunity generated. Mice infected with few parasites did not develop high levels of parasitemia and produced a predominant Th1 response, whereas mice infected with larger numbers of parasites developed substantial parasitemia and predominant Th2 responses. These observations were anticipated within our framework. We further anticipated that infection with few parasites would generate not only predominant Th1 responses but eventually Th1 imprints, and so protection against a challenge with a high number of parasites that causes progressive parasitemia in naive mice. Once this expectation was confirmed, we were in a position to consider how to develop a strategy of vaccination that is effective in a genetically diverse population. As delineated above, we came up with a strategy of vaccinating with an ultra low number of live, attenuated organisms.

In looking at current views, it appears there are two clear but contradictory positions as to how effective vaccination against HIV might be achieved. Most suggest that successful vaccination should guarantee upon natural infection both cell-mediated immunity, in the form of CTL, and antibody, particularly in the form of *neutralizing antibody.* The neutralizing antibody would bind to newly synthesized virus released from infected cells, preventing infection of new target cells by the virus and leading to virus removal. It should be noted that neutralizing antibody can be generated, but that HIV has a means of generating variants quickly and in a manner that most neutralizing antibody, effective against the variant that induces it, is ineffective against some of the variants that therefore become prominent.

The contrasting view is that vaccination must aim to guarantee a predominant Th1 response associated with potent CTL generation upon infection by HIV-1. I share this view for two reasons. First, there are multiply exposed and infected individuals who produce predominant and stable Th1, CTL responses, and who are symptom free. Some female sex workers in Nairobi belong to this group. It is additionally remarkable that they do not seroconvert, and remain healthy over many years despite being exposed on different occasions to major viral variants, or clades of the virus, through their sexual work. Second, all individuals infected with HIV are relatively healthy during their honeymoon period, when their immune system initially mounts a predominant Th1, CTL response, and for a period after they seroconvert, before they start going through the progressive stages of disease. Both these sets of observations strongly support the idea that a predominant Th1, CTL response is protective. Why then not aim for having the best of both worlds, as most suggest? It is an assumption that this is possible, and I feel it is highly questionable. We have seen repeatedly that distinct CD4 T cell subsets tend to inhibit each other's generation, and often each other's effector activities. It may well not be possible to have the best of both worlds, involving a mixed Th1/Th2 response with optimal CTL immunity and the optimal production of neutralizing antibody. It is reasonable to conclude, from observations made in the natural situations described above in which effective resistance against the virus is clearly apparent, that predominant Th1, CTL responses are protective. Further, we can employ this proposition as the starting point for imagining how effective vaccination might be achieved.

Here I describe and outline the reasons underlying my preferred, but untested, strategy to provide universally effective vaccination against HIV. This strategy is based upon similar principles as the one described above to achieve universally efficacious vaccination against tuberculosis, namely the ultra-low dose, BCG vaccination strategy. I recall the essence of this strategy before outlining how additional considerations might lead to effective vaccination against AIDS.

I have proposed in Chapter 7 that administration of low numbers of BCG can lead to universal Th1 imprints and so to resistance against tuberculosis. The number of BCG administered would be so low that it does not induce antibody in anyone, but in time generates Th1 imprints in all.

An important feature of this scheme is that it allows the generation of a BCG-specific Th1 imprint in genetically diverse individuals. Critically, this strategy relies on BCG being a slowly replicating entity. I suggest for similar reasons that it will be most helpful to employ a slowly replicating entity to vaccinate against HIV to achieve universally effective vaccination. In addition, vaccination with an attenuated form of HIV would be unacceptable, as virologists are unlikely to be able to guarantee that the attenuated strain would not revert to become virulent under any circumstances. However, BCG is the vaccine that has been used in humans worldwide more frequently than any other, with the most marginal of side effects. Moreover, people have inserted single HIV genes into BCG in a manner that allows them to be expressed and have employed such BCG vectors to raise immunity against the protein coded for by the inserted gene. Three further considerations and observations make me enthusiastic that BCG vectors expressing HIV genes might be ideal platforms for vaccinating against HIV.

First, HIV has only a few genes, and it is possible to put single HIV-1 genes into different BCG vectors and immunize with a mixture of these vectors. This seems to be a safe way of immunizing against a series of HIV antigens through employing entities that replicate slowly, namely the BCG vectors. Second, in the early 2000s, a graduate student named

Carl Power examined how the Th1/Th2 phenotype of the immune response against the protein, encoded by the vector and referred to as P_V, depended on the number of vector-carrying BCG employed for infection. Infection with low numbers of these BCG vectors resulted in a Th1 response to both P_V and BCG, whereas infection with a higher number resulted in a mixed Th1/Th2 immune response to both P_V and BCG. Thus, infecting with a low number of a mixture of BCG vectors is expected to result in a Th1 response to the HIV-1 encoded antigens. Third, we anticipate that infection with ultra-low numbers of an appropriate mixture of BCG vectors will result in a Th1 response and a Th1-imprint against the HIV antigens in all individuals infected. It seems to me this strategy is promising for achieving effective vaccination against AIDS and is worth testing.

Immunotherapy of AIDS at early stages of the disease

Might it be possible to reverse AIDS progression shortly after seroconversion occurs and before the disease has become advanced? I should like to propose a strategy based upon a consideration of three sets of observations.

First, mixed human Th1/Th2 responses against *Leishmania donovani* can be modulated backwards to a predominant Th1 mode in visceral leishmaniasis patients by reducing parasite load. This reduction can be achieved by administering anti-parasite drugs over a period of three weeks, as previously described. I anticipate that a similar and appropriate lowering of the viral burden by administering anti-viral drugs early in the disease, shortly after seroconversion and when there is a significant but modest Th2 component to the immune response, could modulate the immune response of the infected individual to a predominant Th1 state. This is the state associated with the honeymoon period and present in resistant sex workers. Such a modulation would thus reverse the progression of AIDS. Furthermore, the return to such a state, associated with a low viral load, would also make the individual less infectious to others.

Second, this line of reasoning is supported by an examination of the nature of the immune response in seropositive, HIV-1-infected individuals whose disease does not progress, or does so slowly. This examination shows that the IgG anti-HIV-1 antibodies of such individuals are predominantly of the IgG_2 isotype, associated with a predominant Th1 anti-HIV-1 response. This finding again leads to the conclusion that stable, predominant Th1 responses are associated with effective containment of HIV-1 and so protect against AIDS.

Third, some of the findings on treating AIDS patients with anti-viral drugs strike me as highly interesting in view of the observations just described. HIV belongs to a group of viruses known as retroviruses, and drugs that act against them are called anti-retroviral drugs. Patients are usually given a course of treatment involving the administration of multiple anti-retroviral drugs. When this course is designed to dramatically reduce the retroviral load, it is referred to as *highly active anti-retroviral therapy (HAART)*. Sometimes there are sufficiently strong side effects of such treatments that the physician and patient decide to interrupt it. In this case, there is usually a rapid and dramatic increase in detectable virus, known as *viral rebound,* indicating that protective immunity has most probably been undermined by HAART. In addition, it has been reported that in some individuals in which HAART has been interrupted a few times, so the immune system is exposed intermittently to more HIV virus, that further interruption of HAART therapy does not result in viral rebound. There is an interesting interpretation of these observations that, if correct, has therapeutic implications.

Consider the fact, already described, that giving anti-parasite drugs to visceral leishmasniasis patients for three weeks modulates the anti-parasite response from a mixed Th1/Th2 to a Th1 mode. The shortest treatment that is effective was found by trial and error, as the drugs have significant side effects. Consider what might have happened had the drug been less toxic, resulting in prolonged treatment. The level of parasitemia, after three weeks of drug treatment, is such that the patient now mounts a cell-mediated, Th1 response. Further drug treatment is likely to further reduce the load of parasite antigens and, if this reduction is too drastic, it would be insufficient to sustain the cell-mediated, Th1 response. In this case, cessation of treatment would result in parasite rebound. Interrupted therapy might sustain the cell-mediated response and so sometimes result in lack of parasite rebound due to the sustained generation of a cell-mediated response. Our further studies, described in *Rediscovering,* support this proposal.

We anticipate that at the start of HAART treatment and the consequent drop in viral load, the patient's immune response will be modulated from a mixed Th1/Th2 to a predominant Th1 mode. Once this has occurred, the prevalence of antibodies of the IgG_2 isotype will be greater than antibodies of the IgG_1 isotype, as seen in seropositive non-progressors. We anticipate that if HAART therapy is halted at this time, viral rebound will not take place as the patient's Th1 immunity, and associated HIV-specific CTL, will contain the virus. We also suggest that further HAART at this stage, as standardly carried out, would further reduce the viral burden and undermine the sustained generation of the protective CTL, Th1 response. Monitoring of the IgG isotypes among anti-HIV IgG antibodies can be used to ensure the response remains in a predominant Th1 mode and, if the response starts evolving towards a Th2 mode, HAART therapy can be reinstated for a short while. I suggest this might constitute a potential and personalized treatment of AIDS at early stages.

Vaccination against and treatment of tuberculosis
Vaccination

I proposed earlier in this book that the universal establishment of a mycobacterium-specific Th1-imprint upon the immune system would provide universal protection against infection by *Mycobacterium tuberculosis.* The validity of this premise is not obvious. I first address why I think this premise is plausible and worthy of exploration.

A sound approach to vaccination requires an understanding of why immunity succeeds in providing protection under some circumstances and fails under others. Such an understanding requires an identification of the immunological parameters that discriminate the immune states associated with containment of the pathogen from those immunological parameters associated with chronic or progressive disease.

When various parameters of the anti-mycobacterium immune state of tuberculosis patients and healthy infected individuals are compared, some clear but somewhat incomprehensible conclusions can be made. For example, the average level of anti-mycobacterium-specific IgG antibody is clearly higher in patients than in healthy contacts. These studies lead to the conclusion that on average there is a higher level of antibody in patients than in healthy infected individuals. The difficulty in interpreting what this means, in terms of immunity at the level of the individual, is that this tendency is only apparent when the immune parameters of many patients are compared with those of many healthy infected individuals. The level of IgG antibody in many individual patients, however, is actually considerably lower than in most healthy contacts! There appeared to be no way of interpreting these observations in a simple and plausible manner.

Juthika Menon, who developed the low dose vaccination strategy in the murine model of cutaneous leishmaniasis during her PhD studies, later tried to define immune parameters that distinguish the immune state of tuberculosis patients from those of healthy infected individuals. These studies were frustrating for quite some time. Andrew Judd, a doctor friend of mine, had many tuberculosis patients in his practice. It seemed to me his anti-mycobacterium immune state must represent a state of exceptional resistance. Looking at his immune parameters, I noticed that the level of IgG_2 antibody was much greater than the level of IgG_1 antibody, representing a predominant Th1 response. I suggested one day to Juthika that she calculate this IgG_1/IgG_2 ratio for the anti-mycobacterial antibody of all the patients and of all the healthy infected individuals she had studied.

Juthika showed me the results of the calculations the next day. The value of this ratio for healthy contacts varied from 0.001 to 0.3, and for tuberculosis patients from 0.001 to 100. It seemed evident that the qualitative nature of the immune response in tuberculosis patients with an IgG_1/IgG_2 ratio above 1 was different from that of healthy contacts. From what we already knew about the role of Th1 and Th2 cells in the production of different isotypes of IgG antibody, the immune response of these tuberculosis patients had a greater Th2 component than that present in healthy contacts. We came to refer in our discussions to these patients as having *type 2 tuberculosis.* We envisaged that their anti-mycobacterial response had a significant and detrimental Th2 component.

What could be the cause of failure of the immune response in tuberculosis patients that had a similar IgG_1/IgG_2 ratio as healthy contacts? These individuals generate an immune response that is qualitatively similar to that generated in healthy contacts. Two considerations led me to propose that the immune response in these patients was too weak to contain the infection. In this case, the mycobacterial burden would increase and substantial lesions would form. We refer to this hypothetical form of tuberculosis as *type 1 tuberculosis*.

One consideration favoring this possibility arose from our observations in the mouse model of human cutaneous leishmanisasis. When we inject a million *Leishmania major* parasites subcutaneously into the footpad of mice belonging to resistant strains, a big lesion develops over some weeks before it resolves. It takes considerable time to develop an immune response when a naive mouse is injected with so many living parasites, and during this time, the parasites obviously increase in number and a substantial lesion forms. I thought the appearance of such massive lesions in the lung, caused by a similar immunological situation in an individual whose lung harbored *Mycobacterium tuberculosis*, might lead to pathological symptoms apparent as *type 1 tuberculosis*.

This possibility also provided an explanation for some facts that had troubled me for years. It appears to be a rather general rule that, under natural situations, predominant Th1 responses are either stable or the response evolves into one with a mixed Th1/Th2 phenotype, but that modulation rarely if ever occurs in the opposite direction under natural conditions. It is presumably because of this direction in the evolution of the immune response that people with type 2 tuberculosis become more ill with time and do not recover.

In addition, it is well known that, before the advent of antibiotics, a substantial fraction of tuberculosis patients sent to be treated in sanatoria spontaneously self-cured. It had always seemed to me that such spontaneous cures were not readily explicable on the hypothesis that these patients were ill because their immune response had a debilitating Th2 component. However, mice resistant to *Leishmania major* and given a large parasite challenge form large lesions that resolve with time. The lesion grows until the Th1 immune response achieves an intensity where it kills the pathogen at a rate faster than the rate at which the number of pathogens is increasing through replication.

A reading of the literature, with the hypothesis in mind that there were two types of tuberculosis associated with distinct types of failure by the immune system, was no longer depressing. I suggest that the establishment of mycobacterium-specific Th1 imprints will protect individuals who otherwise would develop either type 1 or type 2 tuberculosis. Individuals with a tendency to develop type 1 tuberculosis would, following vaccination, presumably make a more rapid and stronger Th1 response.

Treatment

It is worth considering the implications for treatment of the hypothesis that there are two distinct types of tuberculosis. People are optimally treated these days by administering a combination of antibiotics that kills the pathogen. Treatment is required for months and is often logistically difficult to complete, thereby giving rise to drug-resistant strains of *M tuberculosis*. The treatment directly targets the bacterial pathogen and there is no attempt to harness the patient's own protective immunity. Therapies that attempt to optimally harness the patient's protective immunity might lead to shorter and more effective treatment. Moreover, it seems likely that optimal treatment will be different in people ill with type 1 and type 2 tuberculosis.

We have outlined above how AIDS patients may be treatable shortly after seroconversion by administering HAART for a relatively short time to optimally modulate the anti-HIV response back from a mixed Th1/Th2 to a predominant Th1 mode. The same strategy should be effective in patients with type 2 tuberculosis. Once the Th2 component has become minor, as assessed by a small IgG_1/IgG_2 ratio, it may be possible to stop antibiotic treatment and allow the patient's immune system to take care of the infection. This would mean the proposed treatment is similar in form to that now employed to treat visceral leishmaniasis. It would include monitoring of the Th1/Th2 phenotype of the response against *M tuberculosis* by the IgG isotype methodology to realize a personalized and effective therapy.

In thinking about an optimal treatment for type 1 tuberculosis, I find it helpful to reflect on why there might be two forms of the disease, on the basis of what we know about how immune responses are regulated.

We have seen in Chapter 7 that different strains of mice respond differently to a standard challenge of *Leishmania major*. It is useful to define a transition number of parasites for each mouse strain. When considering the nature of immune responses following infection by a parasite strain by a given route in mice belonging to different strains, infection below the transition number results in a stable, long term Th1 response; with a number above the transition number in a response with a substantial Th2 component. We have seen in Chapter 7 that the transition number for *Leishmania major* can differ by a factor of 10^5 in different mouse strains. Consider the consequences of infection by *Mycobacterium tuberculosis* in people for whom the transition number for such an infection also varies greatly.

Remember in addition that 90-95% of people infected with *Mycobacterium tuberculosis* do not become ill, presumably because they generate a substantial Th1 response soon, perhaps within a month or so after infection, and so can contain the pathogen at a low level. We suggest that the immune response in some patients with type 2 tuberculosis cannot contain the infection due to the development of a substantial Th2 component of the response. Such individuals are likely to be those who have a low transition number for *M tuberculosis* and who are infected with a relatively large inoculum of bacteria. In this case, they would not develop a sustained and stable Th1 response, but one with a significant and detrimental Th2 component.

Most infected individuals, about 95%, make a sufficiently strong Th1 response sufficiently rapidly that the infection is contained at a low level not associated with pathology. We have argued that in some individuals, in contrast, the Th1 response is made too slowly to initially contain the infection. The initial infection spreads in an unrestrained manner and becomes substantial. The substantial Th1 response that eventually develops causes pathology by the formation, as we have seen, of granulomas in the case of lung infections. What types of people would develop such weak responses and so become type 1 tuberculosis patients? It seems highly likely that these would be individuals with high transition numbers compared to the large majority of individuals that make a substantial Th1 response sufficiently rapidly to contain the infection. Suppose the transition number is 10^4 for such an individual that develops type 1 tuberculosis.

We have seen that the tempo of the response is highly dependent upon the antigen load (Figure 6 of Chapter 4). Thus, such an individual might optimally generate a Th1 response when infected with 5×10^3 mycobacteria. However, if infected with only a hundred mycobacteria, the bacteria will likely grow unimpeded until they reach a level of say 10^3 mycobacteria, when the immune system starts mounting a small response. Consider now an individual with a ten-fold lower transition number; this individual will start producing a response when there is a bacterial load of about a hundred mycobacteria. The inference is that people with type 1 tuberculosis are likely to have relatively high transition numbers. Such patients do not yet have a sufficient Th1 response at the time of diagnosis to contain their mycobacterial burden, unless they have reached a stage where spontaneous remission is about to occur.

In this case, the standard antibiotic treatment now given is expected to reduce bacterial load and so undermine the patient's generation of endogenous and effective immunity against the pathogen. It may well be feasible to develop a strategy of administering mycobacterial antigens in conjunction with antibiotic therapy that leads to much shorter and effective treatment. The consequences of administering such mycobacterial antigen, on the anti-microbacterial immune response, could be monitored, and the amount of antigen given adjusted so that the patient does not develop a substantial Th2 component to their immune response. Such monitoring could be achieving by longitudinally assessing the relative prevalence of IgG_1 and IgG_2 isotypes among anti-mycobacterial IgG antibodies. Such a personalized treatment seems simple. I expect its exploration and development could lead to a much shorter and more effective treatment of type 1 tuberculosis.

Cancer

There are likely diverse reasons why different cancers are not contained by the immune system. Optimal and successful treatment may be aided by an assessment of why immunity has failed to contain the cancer, particularly if the treatment is to be based upon immunotherapy. I describe in *Discovering* the grounds and evidence, gathered by others and ourselves, for thinking there are two major types of failure by the immune system to contain cancer. One is because the cancer is insufficiently immunogenic to raise a strong CTL, Th1 response to contain it. The second is because the cancer generates an immune response associated with a substantial Th2, and perhaps Th3, component that results in a down-regulation of a protective CTL, Th1 response. Our studies in mice suggest that tumor progression often occurs when a substantial Th2 component of the immune response develops. We refer to this possibility as the Th2-Skewing Hypothesis of Tumor Escape. I now consider how these two types of failure might be diagnostically distinguished and how different personalized treatments might accordingly be realized.

It is helpful to start this discussion by referring to a rarely cited paper[26] that I recently came across and that provides a context for the proposals I make.

Human papilloma virus is known to be able to cause cervical cancer in women. Most women do not have substantial antibodies against this virus, but some do. This is usually taken as an indication of infection by the virus. Seropositive women fall into three groups. One group, the patient, has cervical cancer, designated as "Cancer" in Figure 24. Another, the at-risk group, has abnormal cervical structures, often a precursor of cancer, and is

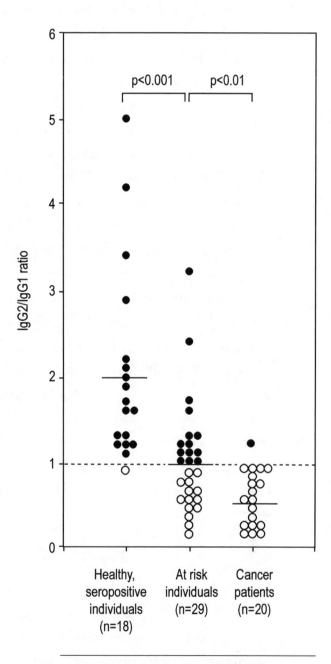

Figure 24. The ratio of the amount of IgG_2 to IgG_1 isotypes among IgG antibodies to a component of papillomavirus among healthy infected (controls), at-risk individuals (at risk), and cancer patients (cancer). Modified from reference 26.

designated as the "at-risk" group of the figure. Finally, the healthy group are not at risk except for the fact that they are seropositive and so presumably are virus-infected, and are designated "control" in the figure.

All patients had a relatively low IgG_2/IgG_1 ratio. At risk individuals had an intermediate IgG_2/IgG_1 ratio, and the healthy individuals had a high IgG_2/IgG_1 ratio. There was surprisingly little overlap in the values of the ratios within a group, particularly between the patients and the healthy controls. These observations are consistent with our knowledge of how the IgG isotype of antibodies produced depends upon the Th1/Th2 nature of the corresponding anti-cancer immune response and the Th2-Skewing Hypothesis of Tumor Escape.

The reason for describing these observations so explicitly is that they illustrate the potential value of being able to ascertain the IgG_2/IgG_1 ratio of anti-cancer antibodies in assessing the Th1/Th2 phenotype of a patient's anti-cancer immune response. Although it is known that many cancer patients mount antibody responses against their disease, simple assessments of the relative levels of IgG_2 and IgG_1 antibodies, and perhaps of other IgG isotypes, against cancer target antigens have not yet been broadly developed. It would not be technically difficult to develop such assays using standard methodologies and employing recombinant cancer antigens against which the immune systems of cancer patients are known to often respond. Our discussion of personalized immunotherapy for cancer patients will be cast in the context that such assays can be developed for different kinds of cancer.

Vaccination against cancer

The implications for strategies of vaccination are straightforward within our framework, and have already been illustrated in considering how to achieve effective vaccination against HIV-1 and *M. tuberculosis*. The low dose vaccination strategy would appear to provide a way of immunizing against cancer, particularly in view of other developments in the field. The investigations by Thierry Boon and his colleagues have defined many tumor-associated antigens recognized by CTL and others have defined the same/similar antigens recognized by antibodies. Such CTL and antibodies are more prevalent in cancer patients than in their healthy counterparts, giving rise to the belief that the immune system provides surveillance against cancer. Tumors of the same type, and even of different types, often share tumor-associated antigens. This finding is important both in trying to prevent cancer by vaccination and in its immunotherapeutic treatment. Again, it would seem most worthwhile to examine whether vaccination with low numbers of BCG vectors, encoding the most frequently expressed tumor-associated antigens, may be effective in preventing cancer. This would protect individuals against making too weak a Th1/CTL response and making a response with a detrimental Th2 component.

Immunotherapy of cancer

Our proposals here are parallel to our proposals of how to treat tuberculosis, and so can be succinct.

Cancer patients with a high IgG_2/IgG_1 antibody ratio against their cancer would be diagnosed as having a predominant and an appropriate Th1/CTL response against their cancer, and a likely cause of failure at containment would be that the response is too weak. It would be appropriate, in this case, to immunize with cancer antigens, against which a response can be detected in the form of conjugates with C, against which the individual has a Th1, cell-mediated imprint. For example, it might be appropriate to immunize with BCG vectors encoding cancer antigens in BCG Th1-imprinted individuals.

Cancer patients with a low IgG_2/IgG_1 antibody ratio against their cancer would be diagnosed as having an ineffective anti-cancer immune response, due to a significant and detrimental Th2 component. The aim of immunotherapy would be to modulate the immune response to an effective Th1/CTL mode.

The two most general ways of treating cancer are to partially remove it, thereby lowering the antigen load, and administering chemicals or treatments that kill dividing cells. It is remarkable that both forms of treatment have the potential for modulating mixed Th1/Th2 responses to a Th1 mode. We have seen in the case of human visceral leishmaniasis that lowering the antigen load modulates a mixed Th1/Th2 response to a Th1 mode.

Irradiation, as well as drugs that kill dividing cells such as cylcophosphamide, are used in cancer treatment. Both these maneuvers have the potential for acting not only on the cancer cells directly, but also on the cells of the immune system, thereby potentially modulating immune responses from a humoral to a cell-mediated mode.

North showed that whole body irradiation of a mouse with a progressively growing and established methylcholanthrene-induced tumor, referred to as meth A fibrosarcoma, could result in tumor regression. He also showed that radiation was effective due to its action on tumor-specific CD4 T cells and not because it directly affected tumor growth. North and Awwad also showed that the administration of cyclophosphamide to an animal, a day before an inoculation of tumor cells that normally resulted in progressive tumor growth, could result in tumor regression. It is known that cyclophosphamide administered a day before tumor implantation means it cannot act to directly inhibit the growth of the tumor cells themselves, as cyclophosphamide is not active a day after being administered. It must act to deplete the recipient of some cell involved in suppressing a protective response. North reconstituted the cyclophosphamide-treated recipient one day after its administration with spleen cells from normal mice to see what cell type was critically ablated and normally responsible for suppressing the protective response. The incriminating cell was found to be a CD4 T cell. It so happens that others had shown that the administration of the same dose of cyclophosphamide, as used by North and given in the same manner, totally inhibits an antibody response to SRBC and that the mice produce a potent DTH response against SRBC instead. All these observations are readily understood in terms of the Threshold Hypothesis and the Th2-Skewing Hypothesis of Tumor Escape, as the generation of Th2 responses requires stronger CD4 T cell cooperation than does the generation of Th1 responses.

It seems to me that, as there is a simple way to longitudinally monitor the Th1/Th2 phenotype of the anti-cancer immune response, treatment could be adjusted to optimally harness the patient's protective response against the cancer. For example, if after two rounds of chemotherapy a high value of the IgG_2/IgG_1 of the anti-cancer IgG antibody indicated the patient had a predominant Th1, CTL response, it might be best to forgo further rounds of chemotherapy, which might undermine the patient's own protective response.

Concluding Remarks

It is understandable if some knowledgeable immunologists, on reading this book, react with the thought that many important topics are missing, both in terms of their description and a discussion of their significance. I agree that this is the case. For example, there is little or no description of NK cells and of $\gamma\delta$ T cells, of the Th17 subset of CD4 T cells, and many other topics besides. These entities are clearly important in some contexts, and they are significant in providing a complete picture of the immune system. They are pertinent for a full consideration of potential immunological interventions in medicine.

My response to such thoughts is two fold. I acknowledge their validity. My purpose in writing this book is not to provide as complete a description of the immune system as is possible at this time. Rather, it is to achieve something incompatible with this purpose. This book represents an exploratory exercise. It examines what limited and minimal observations are necessary, together with careful analysis, to formulate a significant framework that can provide the basis for designing medical interventions. As stated in the preface, it is my view that basic questions are those whose solution provides conceptual platforms for understanding and exploiting the diverse aspects of the immune system. The less information necessary to develop a useful framework and the greater the ability of the framework to account for broad experimental generalizations, the more appealing and valuable it is as a basis for medical intervention. Indeed, the framework developed here was the result of trying to understand the possibilities raised by two basic questions: How is self-nonself discrimination achieved, and how is the class of immunity determined?

A potential insight provided by this framework, if valid, is that understanding the properties and roles of NK cells, $\gamma\delta$ T cells, and the Th17 subset of CD4 T cells, and other topics and entities not delineated here, is not essential to substantially address the two questions considered. Thus, a successful framework reveals the relative importance of different components of the immune system to developing a foundational platform for intervention. In this sense, less is more. Moreover, such a framework as we have developed here, again if valid, allows one to readily incorporate knowledge of those other cells without substantially modifying its foundations.

Some individuals may be uncomfortable with these perspectives. Perhaps they hope that our acquiring more information in the future will inevitably lead to progress and bring with time coherent meaning to all the information we have gathered. I would of course not argue against more information being useful in gaining greater insight. However, the issue I have struggled with in this book is how can we gain better insight given a defined domain of potentially overwhelming information? I am moved to take action by the complexity of the immune system as described in most contemporary textbooks of immunology, and I am inspired by my understanding of the history of science. In the context of immunology, I can only say I am grateful to the spirit of the formulators of the Clonal Selection Theory. They were not daunted by the idea that the immune system is too complex to be comprehensible, nor were they inhibited from making fruitful and empowering speculations.

Glossary

The number after a word indicates the page or pages where
the word is used in a manner to indicate its meaning

References and Notes

1. JH Humphrey and RG White, 1970. *Immunology for Medical Students, Blackwell Scientific Publications,* Oxford and Edinburgh, Third Edition

    Much of the account given here of the history of immunology, up to the early 1900s, is modeled on this classic text.

2. FM Burnet and F Fenner, 1949. *The Production of Antibodies,* McMillan and Co., New York

3. PA Bretscher, 2014. The activation and inactivation of mature CD4 T cells: a case for peripheral self-nonself discrimination. *Scand J Immunol* 79: 348

4. P Ehrlich and J Morgenroth, 1901. Uber Hamolysine: Funfte Mittheilung. Berl Klin. Wschr. English tranlation in The Collected *Papers of Paul Ehrlich, Vol. 1* 1956. London and New York: Pergamon Press pp246-255

5. P Ehrlich, 1909. Ueber den jetzigen Stand der Karzinomforschung. Ned. *Tijdschr Geneeskd* 5:273.

6. L Pauling, 1940. A Theory of the Structure and Process of Formation of Antibodies. *J Am Chem Soc* 62: 2643

7. NK Jerne, 1955. The Natural Selection Theory of Antibody Formation, *Proc Natl Acad Sci* 41: 849

8. P Ehrlich, 1900. On immunity with special reference to cell life (Croonian Lecture) *Proc Roy Soc London,* 66: 424

9. DW Talmage, 1957. Allergy and Immunology. *Ann Rev Med* 8:239

10. J Lederberg, 1959. Genes and antibodies. *Science* 129: 1649

11. FM Burnet, 1959. *The Clonal Selection Theory of Acquired Immunity,* Cambridge University Press

12. S Tonegawa, 19085. The molecules of the immune system, *Scientific American* 253:122-131

13. PA Bretscher and M Cohn, 1968. Minimal Model for the Mechanism of Antibody Induction and Paralysis by Antigen. *Nature* 220: 444

14. PA Bretscher and M Cohn, 1970. A Theory of Self-Nonself Discrimination. *Science* 169: 1042

15. J Yang and M Reth, Receptor Dissociation and B-Cell Activation. 2016. *Current Topics in Microbiology and Immunology* 393: 27

16. C Volkmann, N Brings, M Becker, E Hobeika, J Yang, M Reth, 2016. Molecular requirements of the B cell antigen receptors for sensing monovalent ligands. *The EMBO Journal* 35: 2371

17. CA Janeway, 1993. How the immune system recognizes invaders. *Scientific American* 269:72

18. CA Janeway, Jr., 1989. Approaching the asymptote? Evolution and revolution in immunology. *Cold Spring Barb Symp Quant Bioi* 54 Pt 1, 1

19. P Matzinger, 1994. Tolerance, danger, and the extended family. *Annu Rev Immunol* 12: 991

20. PA Bretscher, 1999. A two step, two signal model for the primary activation of precursor helper T cells. *Proc Natl Acad Sci* 96: 185

21. M Cohn, 2015. Thoughts engendered by Bretscher's Two Step, Two Signal model for a peripheral self-nonself discrimination and the origin of primer effector T-helpers. *Scand J Immunol,* 81: 87-95

22. PA Bretscher, 2015. A conversation with Cohn on the Activation of CD4 T cells *Scand J Immunol* 82: 147

23. PA Bretscher, 1974. Hypothesis: On the control between cell-mediated, IgM and IgG immunity. *Cell Immunol* 13:171

24. PA Bretscher, 2014. On the mechanism determining the Th1/Th2 phenotype of an immune response, and its pertinence to strategies for the prevention, and treatment, of certain infectious diseases. *Scan J Immunol* 79: 361

25. PA Bretscher, 2016. *Reconsidering the Immune System as an Integrated Organ*, FriesenPress.

26. K Matsumoto, H Yoshikawa, T Yasugi, S Nakagawa, et al, 1999. Balance of IgG subclasses toward human papillomavirus type 16 (HPV16) L1-capsids is a possible predictor for regression of HPV16-positive cervical intraepithelial neoplasia. *Bioch Biophys Res Comm* 258: 128

About the Author

Peter Bretscher is a well-known Immunologist. In 1970 he published the Two Signal Model of Lymphocyte Activation, with Melvin Cohn, which provided an explanation for how self-nonself discrimination is realized. This theory has stood the test of time and is a central component, along with the Clonal Selection Theory, of modern immunological thinking. Peter studied physics as an undergraduate at Cambridge University, and then, hoping to find a field offering opportunities for theoretical insight, undertook graduate studies in protein X-ray crystallography in the now famous Cambridge Laboratory of Molecular Biology. He became fascinated at this time by immunology, and was fortunate enough to be able to discuss his early ideas with Francis Crick. This was the beginning of an almost 50 year engagement, during which Peter and his students have made substantial theoretical and experimental contributions to the field.

Peter is the author of numerous scientific papers and recently published *Rediscovering the Immune System as an Integrated Organ*, which looks in detail at the state of immunology today and its relationship to medicine. This book is "an invaluable resource...." (*Foreword Clarion Review*) and a "must read for anyone interested in immunology, a classic book already." Alexandre Corthay, Head of the Tumor Immunology Group, University of Oslo.

Printed in Canada